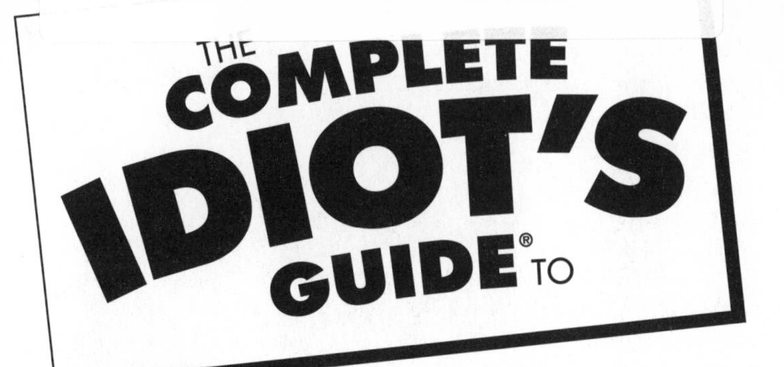


# Extreme Weather

*by Julie Bologna and Christopher K. Passante*

A member of Penguin Group (USA) Inc.

*Julie Bologna would like to dedicate this book to all of those folks who have a fascination with weather and who want to know more about the weather that affects all of us each day.*
*Christopher Passante would like to dedicate this book to his father, Orrie Passante Jr.*

**ALPHA BOOKS**

Published by the Penguin Group

Penguin Group (USA) Inc., 375 Hudson Street, New York, New York 10014, U.S.A.

Penguin Group (Canada), 10 Alcorn Avenue, Toronto, Ontario, Canada M4V 3B2 (a division of Pearson Penguin Canada Inc.)

Penguin Books Ltd, 80 Strand, London WC2R 0RL, England

Penguin Ireland, 25 St Stephen's Green, Dublin 2, Ireland (a division of Penguin Books Ltd)

Penguin Group (Australia), 250 Camberwell Road, Camberwell, Victoria 3124, Australia (a division of Pearson Australia Group Pty Ltd)

Penguin Books India Pvt Ltd, 11 Community Centre, Panchsheel Park, New Delhi—110 017, India

Penguin Group (NZ), cnr Airborne and Rosedale Roads, Albany, Auckland 1310, New Zealand (a division of Pearson New Zealand Ltd)

Penguin Books (South Africa) (Pty) Ltd, 24 Sturdee Avenue, Rosebank, Johannesburg 2196, South Africa

Penguin Books Ltd, Registered Offices: 80 Strand, London WC2R 0RL, England



International Standard Book Number: 1-59257-553-6
Library of Congress Catalog Card Number: 2006925730

08 07 06 8 7 6 5 4 3 2 1

Interpretation of the printing code: The rightmost number of the first series of numbers is the year of the book's printing; the rightmost number of the second series of numbers is the number of the book's printing. For example, a printing code of 06-1 shows that the first printing occurred in 2006.

*Printed in the United States of America*



**Publisher:** *Marie Butler-Knight*
**Editorial Director:** *Mike Sanders*
**Managing Editor:** *Billy Fields*
**Executive Editor:** *Randy Ladenheim-Gil*
**Development Editor:** *Michael Thomas*
**Senior Production Editor:** *Janette Lynn*
**Copy Editor:** *Emily Garner*
**Cartoonist:** *Shannon Wheeler*
**Cover Designer:** *Bill Thomas*
**Book Designers:** *Trina Wurst/Kurt Owens*
**Indexer:** *Heather McNeill*
**Layout:** *Ayanna Lacey*
**Proofreader:** *John Etchison*

# Contents at a Glance

## Appendixes

# Contents

# Introduction

Every day it seems as if there is a new report or article published regarding changes in weather, climate, global warming, or a new storm that ravaged some remote or populated corner of the earth.

A hotter planet, a stormier ocean, more violent winters, more frequent hurricanes … and never have we been more tuned in!

Is it that weather is more severe or that we have more access to weather news, what with 24-hour cable TV channels dedicated to weather, a stronger emphasis in newspapers, magazines, and news stations, and the Internet?

The answer is a categorical yes.

But why? Certainly technology has allowed weather forecasters and scientists to broadcast the latest weather information within seconds, but the technology of forecasting and the unilateral communication between counties—even the most remote—has never been better.

And people—readers, viewers, and everyday people—have never been more aware of their surroundings.

Now, folks can watch a hurricane like 2005's Katrina raging through New Orleans from their living rooms. They watch footage of flood damage along the Mississippi. They can see the damaging effects of a warmer climate on Canadian pine trees. It's all so readily available.

Does that mean it only appears weather has been more extreme? Not necessarily. The latest data show that more violent and frequent storms have plagued the globe, and may not be leaving anytime soon. So now is a great time to be engaged in why weather is important, not only where you live, but all over the planet. An extreme dust storm in Mongolia could have a direct effect on your morning walk in downtown Chicago. A warmer southern Atlantic Ocean could mean a colder winter in Europe. And, of course, deforestation along the Amazon could have a profound effect on your air quality in Miami.

## How This Book Is Organized

This book is divided into five parts:

In **Part 1, "Extreme Weather All Around Us,"** we find out what exactly extreme weather looks like, what causes it, how we forecast for it, and what kind of destruction it can do. There are many components to heavy weather—Mother Nature might be the driving force, but humans are having a larger effect on not only how Earth sustains storms but how strong and frequent the storms become.

**Part 2, "The Answer Is Blowing in the Wind,"** covers wind-driven storms, arguably the most destructive forces of nature on the planet. From hurricanes to tornadoes to desert sandstorms, there is no match for the unpredictability and sheer size and strength of extreme wind storms.

In **Part 3, "A Reason for the Freezin'," **we'll focus on blinding snow, debilitating ice, stinging sleet, and extreme cold. Blizzards can bury entire regions in a matter of a few hours, bringing death and destruction in the form of frozen precipitation and extreme cold.

**Part 4, "Keeping Your Head Above Water,"** is all about torrential rains and severe flooding. When the rain comes in many parts of the world, death and injury, as well as destruction of vegetation, homes, and animals, can be the result. We'll take you inside the storm to some of the worst places and cases around the world.

**Part 5, "The Heat Is On,"** covers heat waves, slice-it-with-a-knife humidity, and heat-related illness and storms. We'll show you some of the hottest, humid places on the planet, and how to keep your cool!

In addition, there are three appendixes with information on weather lingo; record extreme storms including heat, snow, cold and rain, and global hurricane data; and good further reading.

Finally, this book also comes with a special CD-ROM, which will take you around the globe with an in-depth visual look at some of the most extreme weather on the planet.

## Extras

Throughout the book, you'll also encounter tidbits of information that have been highlighted by friendly icons. Here's what to expect:

**def•i•ni•tion**

These sidebars define words that may be unfamiliar to you.

**Storm Stats**

Here you'll discover interesting trivia, including record weather and freak events, with an eye toward the extreme.

**Inside the Storm**

These sidebars cover the science of storms, the ins and outs of predicting storms, how storms "work," and how technology and information help us understand extreme weather.

**Eye of the Storm**

In these sidebars you'll look right into the storm, its power and destruction, discover storm-prone places, and read interesting facts and trivia about what storms do.

## In This Edition: A CD!

Want to look straight into the eye of the storm? Check out the Extreme Weather CD inside this book. See the devastation brought on by hurricanes, tornadoes, floods, and ice- and snowstorms.

We'll also show you satellite images and how weather forms. The CD will play on any modern computer—PC or Mac! And you can navigate the CD with easy-to-follow directions and narration.

The awesome power of Extreme Weather is just a mouse click away!

## Acknowledgments

Julie would like to thank the following people for their part in helping with this book. To my husband, Phillip Dodd, for his love and encouragement. To my parents, Madeline and Domenic Bologna, for teaching me to follow my dreams. To Kimberly Lionetti and Chris Passante for giving me the opportunity to be a part of this book. To my co-worker and friend, Jeff Jamison, for editing my weather copy. To Lizzie, Lester, Erma, and baby-boy-to-be, who were all there with me in the early morning and late night hours when I worked on the book. Finally, to God for all of the blessings that he has given me in my life.

Christopher thanks all those who've helped out in every facet of this book: Bob Sofaly for his photo expertise; Ken Hawkins for his mastery of graphic illustration; Morgan Bonner for his brilliant work on the CD-ROM; meteorologist and journalist Julie Bologna for her vast weather knowledge and expertise on this project; Kimberly Lionetti at Book Ends for connecting me with this project and for her perseverance; Executive Editor Randy Ladenheim-Gil for her guidance and patiently answering every question under the sun; Development Editor Michael Thomas because every writer needs a great editor; Senior Production Editor Janette Lynn for her enthusiasm and direction; Copy Editor Emily Garner for some of the smartest comments and observations through the editing process; and all the good folks at Penguin and Alpha Books who have helped with this project. I also would like to thank my family for their unconditional love, support, and excitement, and God for these gifts.

But mostly I would like to thank my bride, Robyn, for encouraging me to follow my dreams and helping me to achieve them.

## Special Thanks to the Graphic Illustrator

A special thanks goes out to our graphic illustrator, who helped to bring extreme weather to life. All graphic figures in this book were drawn by Ken Hawkins, our resident genius of an illustrator. Ken is a graphics editor at *The Charleston Post & Courier* and had been graphics editor at *The Beaufort Gazette*, both in South Carolina.

## Special Thanks to the Technical Reviewer

*The Complete Idiot's Guide to Extreme Weather* was reviewed by an expert who double-checked the accuracy of what you'll learn here, to help us ensure that this book gives you everything you need to know about extreme weather. Special thanks are extended to Jeff Jamison.

Before arriving at CBS 11 News in Dallas/Fort Worth, Texas, Jeff was a meteorologist at KTBS-TV in Shreveport, Louisiana, where he covered Hurricanes Lili and Isadore along the Louisiana coast. Jeff was the only meteorologist in North Texas to fly into the eye of Hurricane Rita before it made landfall in September, 2005. In 2006, he was awarded the Certified Broadcast Meteorology Seal of Approval, the first in the Dallas/Fort Worth market.

Jeff graduated from Texas A&M University with a degree in Meteorology. He also is an active member of both the American Meteorological Society and the National Weather Association.

## Trademarks

# Part 1

# Extreme Weather All Around Us

Sure, we all see the damage that extreme storms cause, but how they form can be somewhat of a guessing game. Meteorologists have a good lead on why most weather goes bad, but not always. And it's not just Mother Nature who controls the spin of a storm. Man also has a hand in some of the heavy weather and climate changes Planet Earth has been witness to.

In this part of the book, we'll look at what extreme weather is and how it forms, as well as how meteorologists forecast weather with some pretty nifty tools. Also, we'll check out some global-warming debates and how man battles nature in the extreme weather arena.

Chapter 1

# Forecast for Disaster

## In This Chapter

- When weather goes extreme
- What is and isn't extreme weather
- More people at risk
- Mega-extreme storms

Weather might be all around us, but where extreme weather strikes is the last place you want to be around. Each year, severe storms kill and injure tens of thousands of people and cause billions of dollars in damage. Tragedy could be averted with a good dose of common sense and an eye toward the forecast, but severe storms are complicated beasts: Hurricanes take sudden turns, tornadoes spawn on sunny days, sandstorms pummel sailboats fifty miles off the African coast, and freezing rain brings a major metropolitan city to its knees.

Add to the fray that more people than ever are crowding into storm-prone areas like the Southeast and Gulf Coast—practical hurricane magnets and prone to severe flooding—and the result can be cataclysmic. Storm damages on the East Coast in the busy 2004 hurricane season stretched into the billions of dollars and flooding in the low-lying areas on the Gulf Coast is a growing concern as wetlands are traded for waterfront homes.

In this chapter, we'll talk about extreme weather and why we keep finding ourselves in it.

# What's in an Extreme Storm?

Storms of biblical proportions—the fabled forty-day, forty-night flood—are not as common as the run-of-the-mill thunderstorm, blizzard, or tornado. But extreme storms might be more common than you think.

*A sailboat rests on a tree on Bay St. Louis's Beach Boulevard after Hurricane Katrina's powerful storm surge lifted it from its mooring.*

*(Courtesy of National Oceanic and Atmospheric Administration [NOAA].)*

## Hurricanes and Cyclones

Americans have witnessed greater hurricane activity from 1995 to 2005, compared to years past. An average of 7.7 hurricanes and 3.6 major hurricanes have occurred each year since 1995, compared to five hurricanes and 1.5 major hurricanes on average in the 25-year period from 1970 to 1994. The average number of named storms per year from 1995 to 2005 was 13, compared to 8.6 in the 25-year period preceding.

**Eye of the Storm**

In 2005, Hurricane Katrina had a force so intense that it ranks among the most powerful hurricanes to hit American soil in recorded history. Only four U.S. storms in the last 100 years measured greater in sustained winds when they made landfall.

All this talk of the ever-frequent and powerful Atlantic hurricanes may lead you to believe that the southeastern United States is the only place on earth to be menaced by *cyclones*. Not so. Among the most cyclonically active places on the Blue Planet are on the Asian, Indian, and Indonesian coasts. This region is home to some of the most active and strongest storms on earth.

We'll chat about the biggest and baddest cyclonic storms in Chapter 5 and exactly what constitutes the biggest and baddest in terms of lives lost, monetary costs, and intensity. But to punctuate the point, the most severe storm in terms of loss of life on American soil was Hurricane Galveston, so-called in 1900 when storms weren't given male or female names. Galveston killed approximately 8,000 people. In world contrast, however, the Bangladesh cyclone of 1970 claimed 300,000 lives.

**def•i•ni•tion**

A **cyclone** is a depression—an area of lower pressure than the pressure around it. Cyclonic winds blow counterclockwise in the Northern Hemisphere and clockwise in the Southern Hemisphere.

Now, that's big and bad.

## Tornadoes

But if hurricanes are scary, then tornadoes, another type of cyclonic activity, are downright terrifying. Their sheer magnitude combined with their erratic path makes them hard storms to predict.

*A tornado forms near Anadarko, Oklahoma, on May 3, 1999, moments before the F5 Oklahoma City tornado struck.*

*(Courtesy of OAR/ERL/National Severe Storms Laboratory [NSSL].)*

One thing that we can say with absolute certainty is the flatter, the better, for tornado proliferation. We'll get into the nitty-gritty of tornado composition in Chapter 6, but

**Inside the Storm**

Winds inside an F5 tornado—the maximum measurement on the Fujita Scale—can reach a dizzying 318 miles per hour.

the unpredictability of tornadoes has led to countless lives lost and millions of dollars in property damages.

Such was the case in south-central Kansas, eastern Oklahoma, and northern Texas, when more than 70 tornadoes struck the region during the Central Oklahoma Tornado Outbreak of May 3, 1999. The extreme weather event still stands as Oklahoma's largest tornado outbreak ever recorded. More than 40 people in Oklahoma alone were killed in storms that reached a strength of F5, and 675 people were injured. The total damage estimate for the region was $1.2 billion.

How does one measure just how violent a tornado is? Size, number of people killed or injured, or how many tornadoes "broke out" during one storm? We'll get to that in Chapter 5, but check this out: the biggest, fattest tornado ever measured on American soil was in Hallam, Nebraska, on May 22, 2004. At its peak, the F4 twister measured two and a half miles!

## Water, Water Everywhere

Precipitation can be another stormy beast altogether. What the temperature and the wind can do to liquid is on par with some of the strongest heavy weather in the world: anywhere from massive floods to extreme drought, ice and sleet, and *blizzards*, to name a few possibilities.

*Red Cross workers search for victims buried in cars following the February snowfall during the Blizzard of '77 in Buffalo, New York.*

*(Courtesy of the National Weather Service.)*

Major metropolitan cities have been crippled by snow, as was the case on March 12, 1888, in New York City. The weather event, dubbed "The Great White Hurricane," pummeled the East Coast from the Chesapeake Bay area up to Maine.

When it was over, an estimated 400 people were dead and storm damages at the time—1888—were projected at $25 million, or about $500 million today.

Other blizzards certainly have been colder, snowier, or windier, but the triple-threat combination in 1888 puts this storm atop the all-time worst chart. We'll trudge through more snow events in Chapter 10.

**def•i•ni•tion**

A **blizzard** is defined as a storm with winds above 35 miles per hour and snow that limits visibility to a quarter-mile or less for at least three hours. When wind speed eclipses 45 miles per hour, visibility remains a quarter-mile or less, and the temperature hits 10 degrees Fahrenheit or less, then we have a severe blizzard.

What happens when all that snow melts, though? Or when accumulated precipitation has nowhere to go? Flooding is becoming more of a problem as populations move to warmer, flatter, and marshier climates (more on that in a moment). People build homes and businesses on wetlands and pave roads and parking lots over once-permeable surfaces that could accommodate and filter that runoff.

The Great Midwest Flood of 1993 was one of the most devastating floods and one of the greatest natural disasters in American history. More than 50 people died and damages reached $15 billion from the waters that rose and rose again from May to September of 1993 in just about all of the American Midwest. The flood forced the evacuation of tens of thousands of people and destroyed solid ground that may never return to its former state. An estimated 10,000 homes were destroyed and 75 towns were underwater. All told, 15 million acres of farmland were destroyed, barge traffic on the Mississippi and Missouri rivers was halted for two months, and countless bridges on those rivers were washed away.

We'll wade through floods and visit some flood-prone cities in Chapter 14.

## Extreme? Yes. Weather? No

It's a common misconception that much of the naturally caused havoc on Planet Earth is the result of weather. While the gigantic tsunami that battered Indonesia and claimed 225,000 lives in late 2004 was certainly a natural disaster, it is not even close to being *weather.*

**def•i•ni•tion**

**Weather** is the condition of the atmosphere based on temperature, humidity, air pressure, and wind. Extreme weather usually includes some sort of intense heat, cold, wind, rain, snow, or electricity.

Tsunamis are walls of water caused by earthquakes, volcanoes, or landslides under the sea, which displace huge amounts of water and energy from that quick upward movement, forming a wave.

Out in the open ocean, a tsunami might be a three-foot wave traveling at the speed of a jet—about 500 miles per hour. By the time it reaches more shallow coastal water, it slows down, often below 50 miles per hour, but the wave height grows intensely—as much as 100 feet or more. In 1958, a tsunami that struck Lituya Bay in Alaska surged to a high-water mark of 1,640 feet.

Earthquakes such as the one that claimed an estimated 80,000 lives in Pakistan in October 2005 also aren't weather. Again, tectonic plates were to blame.

When you think about the formation of these natural disasters, it becomes clearer what extreme weather is and isn't.

Extreme storms include hurricanes, blizzards, and floods, but in this book, we'll also check out some heavy weather oddities, such as sandstorms, heat waves, deep freezes, and fireballs. Keep in mind that much disaster, too, comes from the after-effects of extreme storms or seasons: California mudslides following torrential rains, ice jams breaking and destroying bridges and roads, snow melting from a particularly severe winter that causes severe flooding downstream ….

Most severe weather can be predicted. But when we choose to ignore the storm warnings, the results are often catastrophic.

For instance, hurricanes are easier to predict than tsunamis. But even with the forecasting technology available, days of warnings, and orders to evacuate, too many people choose to "ride out the storm." In 1999, Category 2 Hurricane Floyd killed 56 people in south Florida alone, despite evacuation orders. In 2005, Hurricane Katrina killed more than 1,000, again despite evacuation efforts. Add to that the thousands injured—including rescue and relief workers—and the exorbitant cost in insurance and rescue aid, and you'll see why it doesn't pay to brave extreme weather.

## Going Coastal: More People in Risky Places

More and more people are moving to areas increasingly prone to severe weather. The beach house on the North Carolina coast or the condo outside of Orlando are fantastic vacation or year-round digs, but they might be fighting with extreme weather more than the ranch house in Cleveland or the bungalow in Pittsburgh.

Dramatic growth has occurred throughout the Southeast, from the Gulf Coast to Florida and into the Carolinas and Virginia. According to 2004 census data, the population of Florida has tripled since 1960, with 17 million people. Between 2000 and 2004, 29 of the 50 fastest-growing counties in America were on the East Coast and Gulf Coast. In the 10-year period from 1990 to 2000, the Sunshine State saw a 23 percent population increase. And it's not stopping anytime soon.

At the same time that folks are packing up the minivan with the kids and Grandma's armoire for warmer climates, hurricanes are packing a bigger punch. From 1995 to 2005, the occurrence of major hurricanes boasting sustained winds greater than 111 miles per hour increased 150 percent. In the Caribbean, where the bulk of the Atlantic hurricanes end up, hurricane production increased by 400 percent in that same time. We'll get into why there's such a marked increase in cyclonic activity in Chapter 5, but for now, it's safe to say that Americans, especially, have never been in harm's way as much as they are today.

**Storm Stats**

Although there is no price tag to be put on human life, there is one on humans' living spaces. The cost of homes, especially in East Coast and Gulf Coast states, has skyrocketed, and that means greater insurance costs. At-risk property has ballooned from about $1.1 trillion in 1980 to $5.5 trillion in 2005.

Americans aren't the only ones moving to the coast. All around the world, folks are heading for the beaches, where rich tourist trades and diverse communities fuel the economies. But, as we have found in coastal America, the risks there are greater, too.

The greatest—and scariest—example of coastal migration may be in the developing countries of the Caribbean. The islands of this region are home to some of the best vacation hot spots on the globe, visited by millions of tourists each year. Their governments have capitalized on their warm, sandy beaches and crystal-blue waters, and their populations have grown exponentially with their tourist economies.

The Caribbean is now the most urbanized island region in the world. From 1995 to 2000, the island nations' urban populations grew on average 1.58 percent annually. That means more people in the Bahamas, Puerto Rico, the Dominican Republic, Cuba, Trinidad, and Tobago. And more people living in harm's way.

Unfortunately, much of these countries' populations are poor—either working class or below—and without suitable homes that can weather even minor storms. It's no wonder that these developing countries contain 90 percent of victims of natural disasters and bear 75 percent of their damages. Also, it's no surprise that of the 25 countries that suffered the greatest number of natural disasters during the 1970s and 1980s, 13 were small island states.

Living on the coast has its perks, but it's pretty crucial to remember that it has its risks, too.

## "The Perfect Storm"

Many of us have read the book or seen the movie, but a good example of some really hard-core extreme weather happens when two or more storms collide. This happened with the "Perfect Storm," largely off the New England and mid-Atlantic coasts, named by the National Oceanic and Atmospheric Administration (NOAA) in October 1991 and later documented in a book and a movie.

When severe weather events combine, the results often are catastrophic. Fortunately, the bulk of these storms are born and die over the vast oceans. With as much as 75 percent of our planet covered by water, the odds are in our favor that two hurricanes won't hook up.

But that's not always the case: the "Perfect Storm" battered the entire East Coast that Halloween weekend.

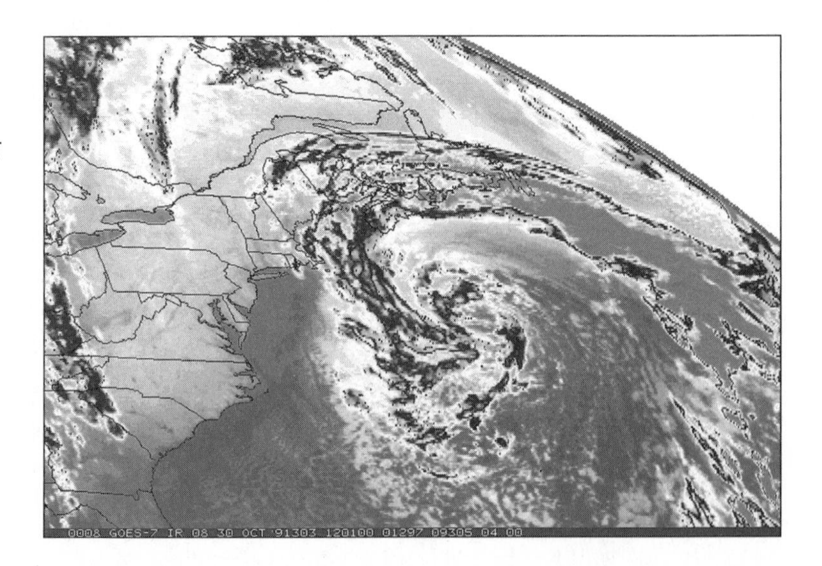


*"The Perfect Storm," an enormous combined low-pressure system, is seen in this satellite image pounding the Atlantic Coast on October 30, 1991.*

*(Courtesy of the National Weather Service.)*

Late fall into early winter is the prime window for the powerful northern Atlantic storms often called nor'easters. Canadian cold fronts take over the midwestern United States at the same time the Atlantic is still clinging to its summer-warmed water. When the two meet, a massive nor'easter tempest forms. Since humans began sailing across the Blue Planet, they've known, feared, and sometimes perished in these terrible storms.

On October 28, 1991, just east of Nova Scotia, a powerful *extratropical* cyclone spun from these late-season weather patterns and was quickly linking up with a cold high-pressure system from Canada. The two patterns met near Nova Scotia at the same time northeast-bound Hurricane Grace, which had formed the day before in the tropics, was barreling up the East Coast. As the late-season Grace combined with the extratropical storm off Nova Scotia on October 29, the "Perfect Storm" formed.

**def•i•ni•tion**

When a cyclone loses its primary energy source from the release of heat and loses its tropical characteristics, it's generally called **extratropical**. Extratropical cyclones can still retain hurricane-force winds.

Winds reached 100 miles per hour over the Atlantic, with ocean waves of 100 feet. By the time the surges struck the New England shore, they were three stories tall. Two hundred homes were destroyed and nine people were killed, including the six-man crew from the Andrea Gail, a Gloucester, Massachusetts, swordfishing vessel.

For five days, from North Carolina on up to the Mid-Atlantic region, as much as 45-mile-per-hour winds howled and tore through coastal areas. The ripple effect of this storm was even greater: flooding from the massive storm reached as far as the Dominican Republic.

Weather history shows scores of nor'easters that have caught people off-guard, and we'll check out many of these storms in this book. For as much forecasting as we can do in this day and age, what two or more storms in close proximity will do often can be unpredictable. The result, many times, is the loss of property or lives.

**Storm Stats**

The "Perfect Storm" wasn't the only nor'easter to catch people off-guard. The Great Atlantic Hurricane of 1944 pounded the Bahamas to New England. Three Navy warships were sunk, killing about 360 crewmen. On land, the deaths were far less, with about 50, but damages adjusted for today's economy totaled $986 million.

## The Least You Need to Know

- Extreme weather primarily includes storms generated by wind, temperature, electricity, and precipitation.
- Extreme weather does not include tectonic or volcanic action (tsunamis, earthquakes, eruptions).
- Record numbers of people are moving to storm-prone coastal areas.
- When two or more storms meet, extreme weather is practically assured.

Chapter 2

# Armageddon? A Stormier Planet

## In This Chapter

- More extreme, more frequent
- Dangers of a warmer planet
- Earth, warmed over
- What lies ahead

It's no fluke: storms in the last decade have become more severe and more frequent. The busiest hurricane season on record was in 2004—at least until 2005, when hurricanes hung in there right through December and into January, two months after the traditional hurricane season ends.

The year 2004 also owns the record for the fourth warmest year, and that autumn was the warmest ever. Five-year droughts in the West, below-average snow packs in the Rockies and Northwest, record-breaking tropical storms in Japan, flooding in Northeast India, severe heat waves in Australia, a rare South Atlantic hurricane in March, massive dust storms in China ….

**Eye of the Storm**

The northeast Pacific averages 27 typhoons a year, with three landfalls. But in 2004, Japan saw a startling 10 typhoon landfalls, beating the record of six in 1990.

End of the world, right? Well, not so fast. Scientists on both sides of the global warming debate are equally adamant whether man or nature is heating up the planet and making for more severe weather. One thing is for certain, though: the planet is a warmer place. In this chapter, we'll look at how global temperature shifts and human activity might affect weather and sometimes contribute to extreme storms.

## More Active Storm Seasons

The granddaddy of all hurricane seasons was 2005, with 27 named tropical storms and 12 hurricanes, including 7 major hurricanes. Never before has a hurricane season run through a complete list of names. With 2005 hurricane names exhausted, storms were named with the Greek alphabet, and in early December Hurricane Epsilon was born, followed by Hurricane Zeta, which spun in the Atlantic Ocean into early January.

Overall, Americans have seen a major increase in hurricanes from 1995 to 2005. An average of 7.7 hurricanes and 3.6 major hurricanes have occurred each year since 1995, compared to 5 hurricanes and 1.5 major hurricanes on average in the 25-year period from 1970 to 1994. The average number of named storms from 1995 to 2005 was 13, compared to 8.6 in the 25-year period preceding.

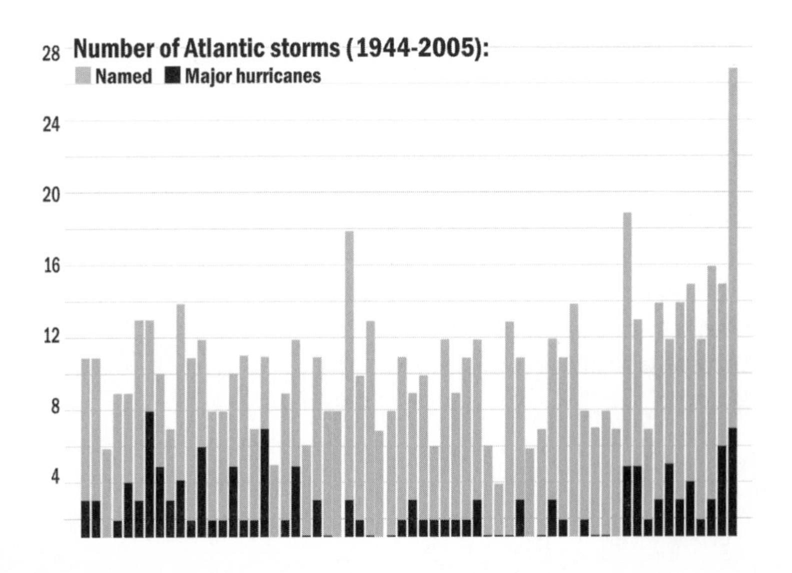

So why the increase in cyclonic activity? Warmer oceans, for one reason. In order for hurricanes to form, temperatures must be near 86 degrees Fahrenheit or higher. This high temperature helps provide the heat and moisture needed to fuel a storm. In the last few years, the water temperatures have reached those numbers early in the hurricane season, making hurricanes easier to form. Plus, the waters have stayed warm later in the hurricane season, making for an overall longer and more active season.

**def•i•ni•tion**

**Ozone** is a colorless gas that acts as a screen for ultraviolet radiation in the upper atmosphere.

Part of the debate on why Earth is getting warmer has to do with whether humans are piercing holes in the *ozone* and killing off natural habitats that disturb ecosystems that could lead to a warmer planet, or if natural cycles are taking place.

We'll dive into the debates later in this chapter, but some scientists also point out 60- to 70-year cycles that alter the strength of currents that disperse heat globally. Some meteorologists believe that one of these cycles began in 1995. There are a few key ingredients that would bring the increase in hurricane activity. One is warmer water temperatures. A continuous flow of upper-level water is pulled from the tropical Atlantic toward the Arctic region, where it cools, sinks, and heads back south deep in the ocean. That warmer water is drawn toward the pole more quickly when the Atlantic conveyor belt is sped up. More speed, more warm water circulating, more hurricanes.

Also, scientists believe *wind shear* is lower, allowing for more hurricanes to develop. And some scientists argue that favorable easterly winds coming off the west coast of Africa can aid in strengthening tropical systems and steering them westward toward the warmer ocean temperatures and lower wind shear in the western Atlantic.

**def•i•ni•tion**

The difference in wind speed and direction between two points is **wind shear,** which can either be vertical or horizontal.

The drop in sea-surface temperatures across the central and eastern tropical Pacific Ocean is known as **La Niña.**

*La Niña* years bring an increase in hurricane activity in the Atlantic basin. When a La Niña weather pattern develops, cooler than normal waters develop in the tropical Pacific and move further westward.

Some meteorologists believe that in addition to having a La Niña event, there would still have to be warmer ocean waters and lower wind shear. So La Niña may not cause a big increase in hurricane formation and activity by itself. Many scientists argue that the active Atlantic hurricane season will continue for many more years and that the active season really began back in 1995.

# Warming Up

Warmer surface temperatures can change overall weather patterns, not just hurricane proliferation. Some scientists worry that a sustained warm-up of a few degrees could create changes similar to those of 10,000 years ago that ended the Ice Age.

## Snowpack

A warmer planet doesn't just affect hurricanes. Snowfall is a major part of the delicate balance of nature. How much snow falls and remains on the ground can mean the difference between a bumper crop or a summer of drought, not to mention a green meadow or the proliferation of dust storms.

Snowfall anomalies are happening, and some are alarmed that they are happening more frequently. The year 2005 saw below-average *snowpacks* throughout the Pacific Northwest and into the eastern Rockies. Drought affected most of the West, including Colorado, Arizona, California, Oregon, Washington, Wyoming, Idaho, and Montana.

> **def•i•ni•tion**
>
> **Snowpack** is naturally packed snow that melts in the warmer months and is used for water resource management.

Concurrently, if the planet is warmer, snowpacks may melt or retreat more toward the poles. Some areas that experience long winters with heavy snow may see shorter winters with less snow and more rainfall. That could lead to more flooding. Plus, exposed terrain (hillsides and mountains that typically have snowpack) that sees increased rainfall can be prone to more erosion. This could change river flows and patterns as well as lake outlines.

## Climate Changes

Just like the currents of the sea, air currents play a strong role in the delicate balance of nature. And there are several influences, but maybe none as prolific as the monsoons.

Monsoons happen when there is a reverse in direction of wind systems. They reverse or change direction each season in several areas of the world, including the southwestern United States. In the winter, the winds along the coast of the Gulf of Mexico frequently come from the north. In summer, they come out of the Gulf and enter the southern United States, bringing up warm, moist air.

If sea surface temperatures are warmer, as they are now, the result could be a hotter, moister summer and even a slightly warmer winter. This could change the typical weather that we experience all over the country. The monsoons could shift farther to the west or east and change a region's seasonal weather pattern.

Some of the worries that exist with the shifting winds and warmer oceans are below-normal rainfall levels and drought, warmer summer temperatures and heat waves, and melting glaciers and snowpacks.

The western United States began 2004 well under normal rainfall averages. A nearly five-year drought plagued parts of the West in the first half of that year. Short-term relief came in the fall, but not enough to offset the rainfall deficit. The Southeast also suffered a five-year drought that ended in 2005. Rainfall averages were far below average, which not only affected water shortages for agriculture, but the lower water table also affected South Carolina's shrimping industry, forcing shrimp to cooler water upriver, where commercial shrimp boats couldn't retrieve them.

**Storm Stats**

The year 2004 was the fourth warmest since 1880 worldwide, 0.97 degrees Fahrenheit above the 1880 through 2003 long-term mean.

The dry conditions also didn't help the timber and brush lands in the Southeast, including Florida, which suffered multiple years of wildfires that consumed hundreds of thousands of acres.

But back to the monsoons. The "real" monsoons are in south and southeast Asia. These shifting winds signal changes in seasons, and drastic changes in weather—wet in the summer and dry in the winter.

Change in the monsoon seasons is something that is keeping many climatologists and scientists on their toes. First off, the seasons themselves seem to be longer, later, wetter, and drier, depending on the seasons. Flooding in Bangladesh, for instance, has been more frequent and severe over the last decade.

**Eye of the Storm**

The worst monsoon in history was in 1983 in Thailand when 10,000 people died and an estimated $400 million in damages were tallied.

Asian monsoons have been slowly gathering strength in the past few centuries, and the warming of the planet is a contributing factor—so concluded the Intergovernmental Panel on Climate Change.

So we've seen some severe weather anomalies around the world. But do humans really have a hand in those freak weather patterns? The way to answer that is to look at both sides of the debate.

Here's what we know for certain: in 2004, the average global temperature for combined ocean and land surfaces was 0.97 degrees Fahrenheit above the long-term mean from 1880 to 2003. The year 1880 just happened to be the beginning of reliable instrumental weather records. Worldwide extremes were felt in each season.

# Global Warming

The *global warming* debate has come into the spotlight since the 1980s, with a renewed conservation movement in the United States and across the globe. Worries over holes in the ozone and the greenhouse effect have even been the subject of novels and made-for-TV movies.

**def•i•ni•tion**

**Global warming** is a rise in the average temperature in the earth's atmosphere, sufficient enough to cause climatic change.

But there has been very serious debate in scientific communities, and many reports have been penned in the halls of the United Nations.

So why the perception that there is no definite conclusion?

We know the planet is a degree warmer, that more storms are resulting from that temperature rise, and that humans can have a profound impact on global health.

Melting polar ice caps, more active hurricane seasons, increased dust storms ….

So what's the argument?

First, let's take a closer look at what global warming is.

## What Causes Global Warming?

So-called greenhouse gases, such as carbon dioxide, which build up in the atmosphere, have increased since pre-industrial days. Whether the earth's atmosphere can accommodate those greenhouse gases is the subject of much debate. One thing is certain, though: the atmosphere has been altered by human activity.

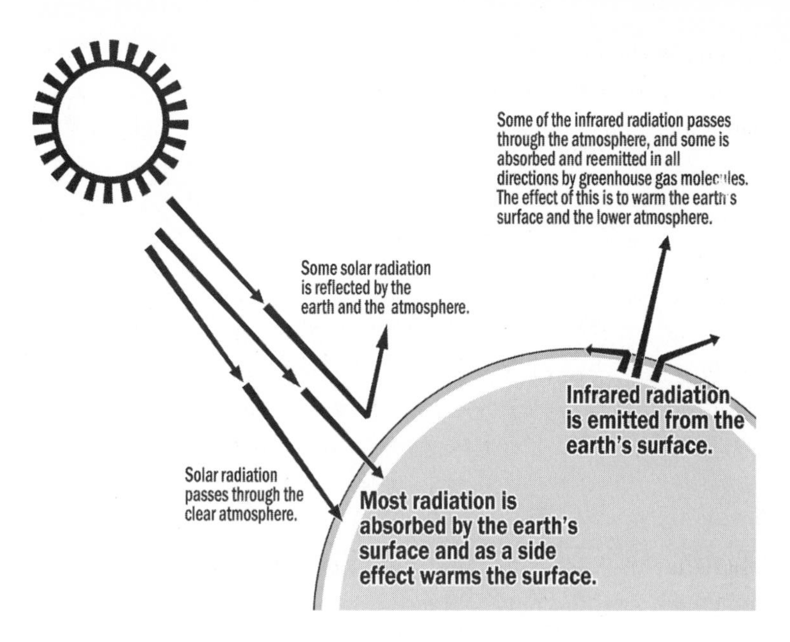


*Solar radiation.*

*(Courtesy of the National Oceanic and Atmospheric Administration [NOAA].)*

The sun is the energy source for the planet, and it has everything to do with the weather and the climate we live in. In a nutshell, the sun sends energy to Earth and Earth radiates it back into space. When greenhouse gases trap that heat energy bouncing back off the planet's surface, the planet gets warmer, just like the sun coming through the glass of a greenhouse.

It's not as if the greenhouse gases normally allow the release of all that heat energy; there is a natural greenhouse effect in place to keep the planet habitable.

But if those gases do block some of that heat, a few degrees of difference could harm or kill valuable ecosystems, as well as alter weather patterns.

**Storm Stats**

Concentrations of carbon dioxide in the atmosphere have increased by nearly 30 percent since the Industrial Revolution.

Use of fossil fuels and other human activity are mostly to blame for increased carbon dioxide concentrations. Fossil fuels are burned in cars, homes, factories, and businesses. In fact, 98 percent of carbon dioxide emissions come from these. Of the total carbon dioxide emissions released in 1997 worldwide, the United States contributed about one fifth of them.

> **Eye of the Storm**
>
> In southeast China during 2004, drinking water supplies were threatened when the worst drought in 50 years gripped the region.

According to national weather experts and scientists, climate change will be even more affected by the greenhouse gas concentrations as the global surface temperatures rise as much as 4.5 degrees in the next 50 years and as much as 10 degrees in the century.

Warmer temperatures mean more precipitation, more flooding, more hurricanes, and more melting of ice caps, which could mean a 2-foot rise of the sea level along most of the U.S. coast.

Many scientists have accepted that these trapped gases have warmed the planet and will be in the atmosphere for centuries. The 1-degree warmer difference since the late nineteenth century has been blamed on melting glaciers, decreased snowpack, and warming of the planet, above and below ground.

Even El Niño events are being studied as possible byproducts of global warming. The debate is whether global warming could have increased the Pacific Ocean's temperature, causing El Niño's effects to be more frequent.

Scientists already are predicting colder winters for much of Europe.

Warmer planet, colder winters? Sure. When the planet is warmer, the ice caps melt. That cold water must go somewhere, so more freshwater enters the ocean, changing the saltwater content. This can slow down and redirect the circulation of the oceans. The warm waters of the Gulf Stream would move along the eastern United States, then turn toward Europe. This might bring about colder weather to Europe and the northeastern United States. These locations might experience more snowstorms and cold weather throughout the year.

## The Other Side of the Debate

Skeptics point out that evidence leads to natural causes being to blame for global warming. A George Marshall Institute report claims that the temperature of the Earth will continue to increase, but a negligible amount in the next century, and maintains that the increase would be insignificant with regard to climate fluctuations. The institute claims its mission is to "encourage the use of sound science in making public policy about important issues for which science and technology are major considerations." The program also "emphasizes

> **Storm Stats**
>
> In 2004, sea ice in the Arctic region was 13 percent below the 10-year mean.

issues in national security and the environment." It may be important to note that the institute sponsors have included American Standard Companies and the Exxon Education Foundation.

The institute claims that humans do not affect global warming and that most of the warming happened pre-1940 when most of the carbon emissions were sent into the atmosphere. The report even questions whether hurricanes, for instance, have increased as a result of global warming.

A United Nations and World Meteorological Organization report concluded that human influence on climate is discernible, and that even a small increase could happen in the next century, causing more frequent and catastrophic storms.

There are many factors that can affect tropical storm and hurricane development. Meteorologists need to look more at the long-term changes over hundreds of years versus only a few years or even a few decades. It's still too soon to know how global warming is affecting hurricane formations, but most scientists agree that if the trend continues, the world could be in bad shape.

## A Delicate Balance

Human factors don't only include blowing holes in the ozone layer or trapping greenhouse gases. Poor farming techniques, overgrazing, over-harvesting forests, destroying wetlands, and killing aquatic life in lakes and rivers could have profound effects on the climate, and, in turn, weather.

Imagine a forest of large-canopy trees, gently blowing in the summer breeze. Walking under that mature-growth forest, the terrain is soft and the air cool—even in the heat of the summer.

Now imagine those trees gone. Nothing but stumps. All of a sudden, that once-soft, cool soil is now parched, cracked, and barren.

And hot. The hotter the ground, the drier the rivers and lakes, the less evaporation from water sources and (less) transpiration (or moisture release) from plants, the warmer the climate.

And we all know what happens then, depending on which side of the debate you're on.

**Storm Stats**

Dry conditions in Alaska in 2004 led to record forest fires that claimed 6.4 million acres.

## The Least You Need to Know

- Planet Earth is in a warming trend that is resulting in stronger and more frequent storms.
- Many dangers exist when a planet warms.
- There are two sides to the global warming debate.
- Conservation methods are needed to maintain a healthy balance.

## Chapter 3

# Tools of the Trade

### In This Chapter

- Caveman forecasting?
- Predicting weather from your living room
- Getting it wrong
- The view from up there

At no other time in history have we been more tuned in to how severe weather affects our planet. At our fingertips are 24-hour weather networks, Internet access to the latest NOAA storm-tracking images, and live television coverage from the eye of a hurricane.

Do a quick Internet search for "weather chat rooms" and you'll find hundreds of venues. People are glued to extreme weather, and it's not hard to see why.

In this chapter, we'll look at how it all began, from early forecasting to today's super-powered satellites and computers. I'll also show you what it's like to fly into a hurricane.

# Primitive Forecasting for Extreme Weather

Keeping up with the weather in earlier times wasn't just a click of the remote control or a hard return on an Internet search engine. But it wasn't all that difficult, either.

Today, as in the day of the caveman, we can walk outside and experience the rain, sun, clouds, cold, heat, wind .... We can determine at that exact moment that it is, in fact, rainy, sunny, cloudy, cold, hot, or windy.

If you, I, or our friend the caveman go outside again tomorrow and find ourselves in one of those same elements, we might conclude that the next hour or few hours—or even days—probably will be the same.

**Inside the Storm**

Primitive forecasting is still practiced today. Arthritic people, for instance, claim they can tell when it will rain based on a pain in their joints. It's true: as barometric pressure changes, it can affect the joints.

Okay, that's pretty basic, right? But consider this: our caveman's great-great-great-great grandson wakes up in the same area that his forefather did on the same day 100 years later. Passed down to him were some simple facts: when the sun is at its hottest for a continued period of time, the streams tend to dry up because it doesn't rain much during this period. Also, when the oceans are warm, great cyclonic storms form. There aren't many hurricanes over cold waters.

For hundreds of years, those facts held firm, and our caveman's ancestors knew that they lived in a climate with a hot, dry season and stormy summers, and that those patterns came around with a fair amount of consistency.

That's forecasting.

## From Aristotle to Old Farmers

Aristotle, the great Greek thinker, dabbled in weather. His work "Meteorology" shows some of the earliest weather predictions on the planet. In 350 B.C., Aristotle wrote: "Now the sun, moving as it does, sets up processes of change and becoming and decay, and by its agency the finest and sweetest water is every day carried up and is dissolved into vapor and rises to the upper region, where it is condensed again by the cold and so returns to the earth."

The philosopher is explaining the hydrologic cycle—the changes from liquid to gas as it moves among atmosphere, land, surface, subsurface, and organisms. Not extreme, but still pretty cool.

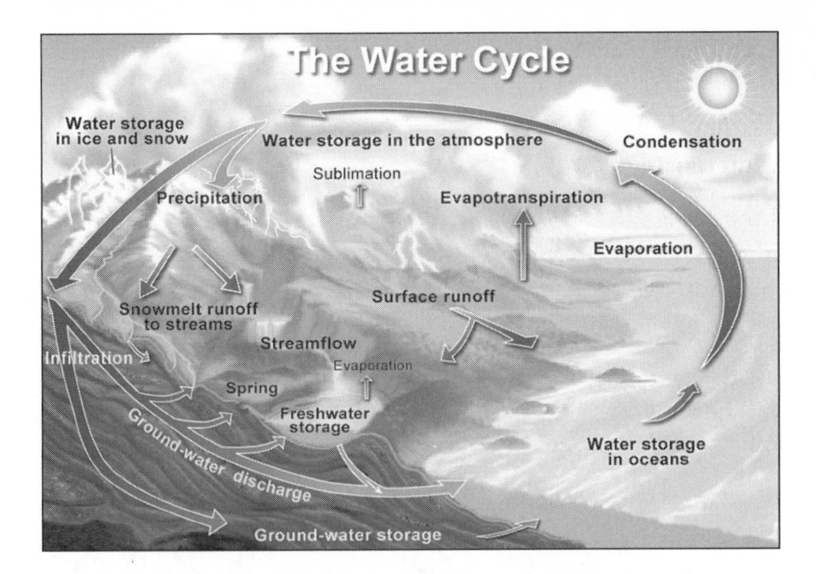


*(Courtesy of the U.S. Geological Survey.)*

And there's a good deal of fact to primitive weather forecasting folklore. Have you ever heard the expression "Red skies in morning, sailor take warning; red skies at night, sailor's delight?" Most of us have, or at least one of the dozens of variations. That's because sailors were among the most accurate and active weather predictors on the globe. Many still are, mostly because their lives sort of depend on it. After all, there were no weather faxes back then.

When early sailors headed out into the Atlantic Ocean, this little saying was widely heeded. For years, sailors before them understood that a big red ball of sun in the morning usually meant rain. The color comes from the vapor in the air, which often prefaces rain. You can try it at home.

*The Old Farmer's Almanac* was another early forecaster—and continues today. People still are perplexed over how this funny little mustard-colored pamphlet can predict weather 18 months in advance. Sure, maybe you don't want to solely rely on the Old Farmer when planning your sail to the Bahamas during hurricane season, but its forecasters have been watching the weather patterns for hundreds of years, and since 1792 have been predicting relatively solid weather—even extreme weather.

**Storm Stats**

Skeptical of *The Old Farmer's Almanac*? Consider this: among its accurately predicted weather events were the July snow of 1816; the 1953 Worcester Tornado; and Hurricane Andrew in 1992. The publication claims a consistency rate of 80 percent.

## Good Gauges

What thrust the primitives into modern forecasting were a couple of pretty critical inventions: the thermometer and barometer.

To this day, folks know how to use a thermometer, and do almost every day, but don't really understand how it works. What? Mercury expands when it's heated? And when it expands it rises in this little glass tube? How scientific is that?

Pretty darn accurate, my friend.

And the barometer? What's up with that?

Without mercury, at least in the early days, we'd be lost.

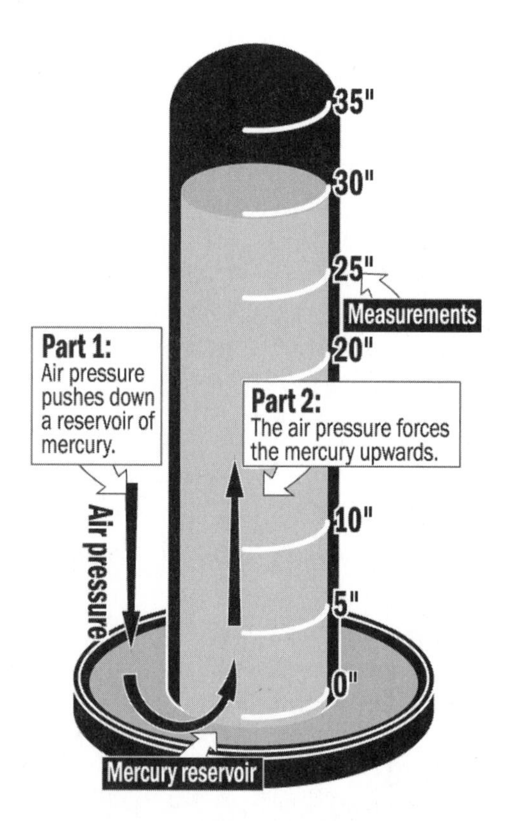


*Barometric pressure.*

*(Courtesy of the National Oceanic and Atmospheric Administration [NOAA].)*

We feel the earth's pressure in our ears when taking off in an airplane or coming down the mountain. Mercury has a knack for reacting to that pressure, and in the figure above we can see how we use it as a gauge. The air pressure relates to its density, which jibes with the air temperature and height above the earth's surface. Air pressure is the weight of the air pressing down on the planet. Weight is the measure of how much force acts on an objective, due to gravity. Since the pressure depends on the amount of air above the point where you are measuring, the pressure falls the higher you go.

We can measure pressure on a barometer, but so what?

Well, foremost is this: low pressure usually equals bad weather, and high pressure usually equals good weather. Extreme low pressure usually means take cover!

Extreme storm centers are usually super low-pressure cells. If you want to forecast, take a glance at a barometer. If the needle is pointing to low, trouble's on the way. We measure pressure in millibars—a unit of atmospheric pressure equal to one thousandth of a bar.

**Inside the Storm**

On June 24, 2003, a violent tornado near Manchester, South Dakota, registered a 100 millibar pressure plunge. Although it's hard to say what an average pressure drop is during a tornado—no complete historical database exists—we can say the bigger the drop, the badder the beast.

You may have heard the old phrase "weather blows from high to low." It means just that. As high-pressure weather sits in one region, and low in another, then the air is going to go from the high to the low, and that difference in air pressure causes the wind. The greater the pressure difference, the windier it will become.

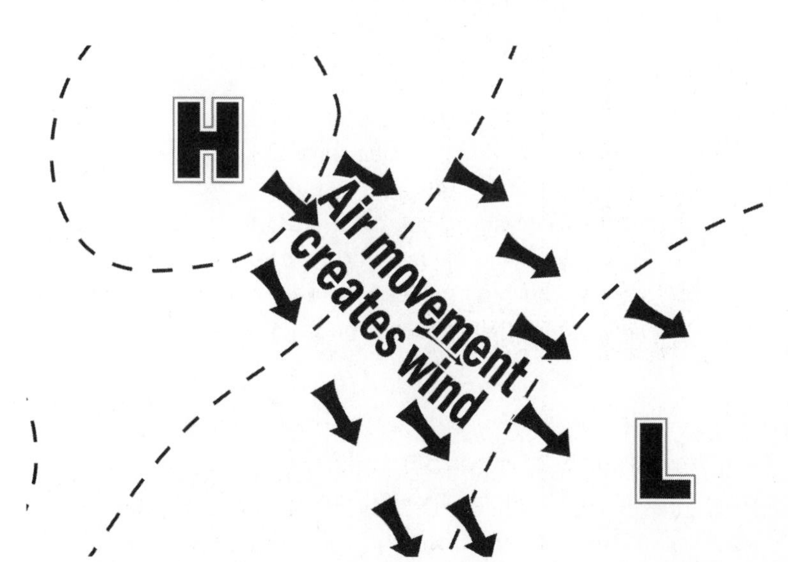


*Air moves from high pressure to low pressure, as shown in this diagram. The movement of the air from high to low causes wind. The greater the difference between the pressure systems, the greater the wind.*

*(Courtesy of the National Oceanic and Atmospheric Administration [NOAA].)*

As air in a high-pressure system sinks toward earth, it becomes warmer. Good high pressure is the difference between a nice warm day on a sunny beach versus the low-pressure system's chilly wind and rain that sends you and everyone else on that beach scrambling for cover. The good news is that high-pressure systems tend to be more stable, and they're bigger, too, so they last a good long time over a good-size area.

In a low-pressure system, the air rises, and the higher it goes, the cooler it gets, like climbing a mountain: you might see snow caps at the summit although it's still balmy and grassy at the base.

As the air rises, clouds usually form and, quite often, it rains.

What's cool about all that? Well, for one, a quick glance at a barometer can tell you all of this. If that dial is heading toward low, bring your umbrella to the parade: it's probably going to rain.

# Modern Forecasting for Extreme Weather

For the moment let's forget all about what color the sky is at night, the history of the last 30 years of weather, and even leave alone the mercury as we step into today's world of extreme weather forecasting.

Not only do meteorologists have hundreds of years of severe weather history to draw from, but they also have a world of high-tech computers and machinery at their fingertips.

With the advent of computer and satellite technology, forecasters can watch storm cells develop from outer space, better understand storm models bounding over vast oceans, and predict to the hour when hurricanes will make landfall.

Meteorologists use information from several computer models to put together short- and long-term forecasts. But many of the computer models have their limits. For instance, some models seem to forecast better for one area of the country than another area. Topography plays a big part in the accuracy of the computer models' forecasts. Also, some models are more accurate in one season than another. So knowing one's forecasting area—do you live in a place with high mountains or in an area with a slight rise or drop in elevation?—will many times affect a meteorologist's forecasts.

## All About Accuracy

A good meteorologist will take the information presented in computer models and tweak it for his or her local area. Being familiar with their topography, river and lake systems, and overall environment can mean the difference between an accurate forecast and a blown one.

Some of the more popular computer models that meteorologists use today are *GFS*, *ETA*, and *NGM*.

**def•i•ni•tion**

There are three major computer models used by today's meteorologists. They vary in range and duration, and each has specific tasks. The three models are **GFS**, **ETA**, and **NGM**.

GFS stands for "Global Forecast System" and takes the place of two former models, the AVN and MRF. It's used for short- and medium-range forecasts as its information runs up to 200+ hours.

ETA was named for a system that is based on mathematical coordinates for a particular topographic area, like mountains or plains. Its range is about 18 miles, so it can give a more specific look at a local area by highlighting changes in elevation or landscapes.

NGM stands for "Nested Grid Model." Also a short-range forecasting model—up to 48 hours—it can look at different weather variables like temperature and wind at different levels of the atmosphere. The atmosphere is sectioned off into grids for different levels. Smaller grids are nested inside larger ones. The smaller the grid, the more computer information that is needed. The NGM's range is about 50 miles.

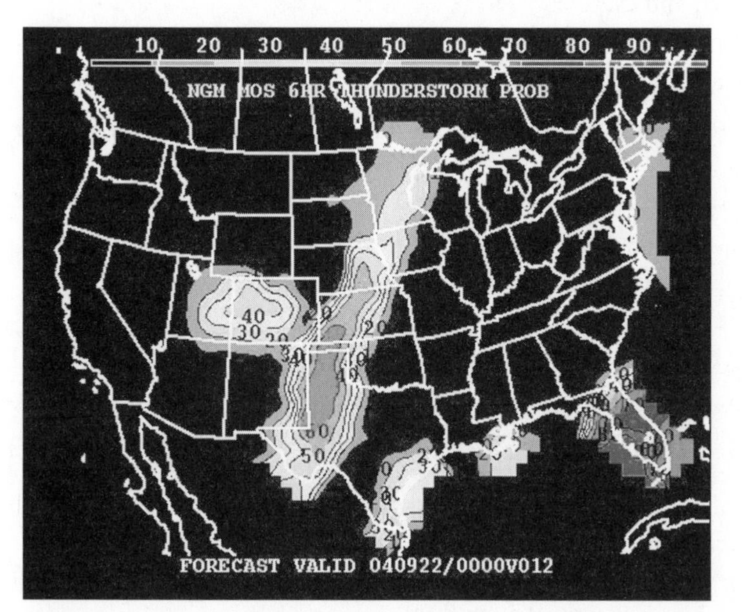


*NGM Computer Model.*

*(Courtesy of the National Oceanic and Atmospheric Administration [NOAA].)*

Beyond history, common sense, and computer models, meteorologists also turn to Doppler radar, satellites, and automated weather stations, especially when severe storms are approaching an area. These tools are critical to "nowcasting"—forecasting up to about six hours.

Using Doppler radar is extremely helpful to forecasters to see directions that storms are moving and the intensity of the storms. Doppler also can be used to detect *bow echoes*, or storms that form a bow shape when moving through an area. These bows often produce severe storms.

**def•i•ni•tion**

**Bow echoes** are radar echoes that are usually associated with severe thunderstorms. The bow refers to the outward bow shape.

A line of severe storms also can take on a characteristic hook shape or "hook echo," which can indicate a tornado. Some Doppler systems have storm-tracking capabilities that look at storm speed and direction and give an estimate of which cities may be affected and at what times. You may have seen this while watching your local weather forecast on TV. It is usually viewed as a long arrow on top of a Doppler radar image that points to the target areas.

But even with this technology, caution must be used. Storm tracking is based on the idea that a storm will continue to move in the same direction and the same speed. However, storms can quickly weaken, speed up, or change directions. So this should be used primarily to get an *idea* of possible storm movement.

Looking at a live Doppler image, meteorologists can zoom down to street level to see which neighborhoods are getting pounded by storms and which areas are not.

Satellite imagery is useful to forecasters because they can detect cloud types that are capable of producing severe weather outbreaks. Tune in to any TV news station during a hurricane watch, and chances are you will see a big swirling menace trekking across the Atlantic, shown over and over. That's a satellite image.

**Storm Stats**

How strong can a tornado get? Typical tornado monitoring devices in the path of a twister can measure up to about an F5 storm. Anything over that and the device is damaged by the storm's powerful winds.

Automated weather stations are set up all over the United States. They automatically give information about temperature, humidity, winds, pressure, and rainfall. These are especially valuable in small towns where weather observers are not available to give a weather report. These stations are usually very accurate; however, there are some problems. Sometimes the stations go down or break. They also require someone to go out and fix them, and sometimes that takes a few days.

*Satellite image.*

*(Courtesy of the National Oceanic and Atmospheric Administration [NOAA].)*

Weather stations can be placed at higher elevations. So when a forecaster uses that information they should take note of where the station is located. For example, if the station is placed in a higher elevation on a mountain, then it will generally show temperatures that are cooler because temperatures usually decrease with height. It will show a temperature that is cooler than the surrounding valley locations. Also, in the case of a temperature inversion, just the opposite happens—temperatures increase with height. It would read a higher temperature than the surrounding valley location.

*National Weather Service forecasters Chris Sohl and Doug Speheger issue warnings from the Weather Forecast office in Norman, Oklahoma, on May 3, 1999, during the Central Oklahoma Tornado Outbreak of the same day.*

*(Courtesy of the National Weather Service.)*

## No Match for Mother Nature

Modern forecasting methods are much more accurate than methods used decades earlier. Yet there still are shortfalls with today's methods. Usually, most meteorologists show a five-day forecast when presenting the weather. The first 48 hours are about 80 percent accurate. The remaining three days should be used just to get an idea about upcoming weather events and future trends. The accuracy of the last three days drops off slightly.

For instance, on Day Three, the forecast calls for mostly sunny skies and a high of 78. On Day Four, thunderstorms are forecast with highs near 80. On Day Five, it is forecast to be mostly cloudy with a high of 70.

With a storm approaching, the forecast area may get so-called above-average precipitation and above-average temperatures (if your average high is only 72). Also, during days Three, Four, and Five, the timing of storms may change. If the storm is forecast to arrive on the fourth day, it could slow down or speed up, causing the five-day forecast to change as you get closer to the end of the forecast period. So, relating days Three, Four, and Five with precipitation and temperature trends is the best way to read that part of the forecast; during days One and Two, you will have a very good idea of what type of storm is approaching, when it should arrive, and how that will affect your temperatures, winds, and sky conditions.

The computer models that forecasters use on a daily basis to interpret the weather can be better in one season and worse in another. Also, the models may be better at forecasting for one area—such as the Southern Plains—but worse at forecasting in the Northeast because of the different terrain and the need to know the difference between rain, snow, or sleet for different elevations.

Overall, modern forecasting has come a long way. It is more accurate now than it has ever been, yet there remain many limitations and variables that technology and humans cannot understand or predict. It seems that in the end, Mother Nature knows the most.

# Missing the Mark

In late August 2004, Hurricane Gaston skipped across the Atlantic as a small hurricane, teased seaboard residents from northern Florida to South Carolina, then parked itself on the Georgia-South Carolina border for a day before setting its sights on Charleston, South Carolina. Emergency management agencies for the three lower states declared evacuations, some voluntary and a few mandatory, based on the hurricane models and chances of a strike that would occur in the pre-dawn hours of August 29.

Residents on Hilton Head Island, South Carolina, on the morning it was expected to make landfall, instead woke up to sunny skies and a light breeze. It was a Sunday, and within a few hours of daylight, the storm was shoving off and folks were literally heading to the beach to enjoy perfectly sunny skies and the spirited surf.

However, historic Charleston, just 75 miles north, took the brunt of the storm, with torrential rain and widespread flooding that crippled the city.

It's not that the forecasters were wrong; even with all the high-tech equipment, computer models, and years of storm predicting, one important variable is that storms, especially hurricanes and tornadoes, are sometimes just unpredictable. Don't trust the weatherman? The emergency operations guys were correct to evacuate based on the data provided. Don't trust the storm? Now we're talking.

**Eye of the Storm**

A record 26 named storms struck the United States in 2005, but the most damaging hurricanes—Dennis, Katrina, Ophelia, Rita, and Wilma—caused their damage in Gulf Coast states.

There still are many challenges ahead for forecasters and technology in forecasting weather, but compared to how they forecast weather 150 years ago, significant improvements have been made.

Even as late as the 1950s, forecasters knew that they were forecasting on shaky ground. Knowing how inexact a science tornado predicting was, the National Weather Bureau banned sending out tornado warnings before 1950 for fear that iffy forecasts would cause mass panics. But after years of documenting and researching weather patterns surrounding tornadoes, predictions became more reliable and in 1950 the Weather Bureau lifted its ban. Probably a good idea ….

It wasn't until 1948 that forecasters got their swagger in predicting tornadoes. Although someone may have claimed they'd forecast a tornado in an area and it happened, the first documented accurate prediction was March 25, 1948, at Tinker Air Force Base in western Oklahoma, when Air Force Capt. Robert Miller and Maj. Ernest Fawbush unlocked the secret of twister predicting. The men saw the similarities in weather patterns that developed from a tornado in their area a few days before March 25.

**Inside the Storm**

By comparing data with similar conditions just five days after a tornado struck, forecasters were able to accurately predict another violent twister and warn residents to take cover. It was the first warning to break a long-held ban.

When they saw those same weather patterns, they went to their superior officer and convinced him to take action. A warning was issued, and on the next day a tornado struck as predicted. In fact, the tornado struck the base.

Folks used to watch cloud formations and lick their finger and hold it in the wind to find out if storms were approaching or if wind directions were changing. While weather forecasting remains an inexact science, it is getting better every day.

## Tracking Severe Storms

You're now familiar with many of the tools that modern and primitive forecasters use and have used to predict weather, storms, and severe storms. Some are exact sciences, right down to the millibar, while others are pretty good educated guesses. Mother Nature is fickle, but we're figuring her out one storm at a time.

Let's take a look at how scientists track an extreme storm. As we've pointed out, there are many different tools forecasters use to forecast a storm, whether it's a Category 3 hurricane born off Africa or a lake-effect snow squall blowing off Lake Ontario.

More and more, we see television news crews taping or going live while hunkered down on beaches as a nasty hurricane approaches. They are relaying information to us viewers—whether we're just a few miles up the coast and need to know whether to evacuate or we're sitting in a comfortable recliner in Fargo watching for entertainment. But they're also plotting the course, checking wave height, measuring wind speeds and shifts, and constantly monitoring data.

Ditto for the so-called tornado chasers. Sure, the blockbuster movie *Twister* glamorized these "mad" scientists and adrenaline junkies, but they also are doing an important job: collecting tornado data, firsthand.

**def•i•ni•tion**

A **storm spotter** observes from a stationary, usually protected, position. A **storm chaser** hunts for tornadoes, usually driving out to where a twister is.

Rob Jensen was believed to be the first *storm chaser* in the 1940s, and David Hoadley began chasing twisters in the mid-1950s. Neil Ward of the National Severe Storm Lab was considered the first storm-chasing scientist, bringing back loads of data to build complex tornado simulations in his laboratory.

Today, there are about 1,000 storm chasers in the United States, but they are predominantly hobbyists. What goes into tracking a storm is a combination of science and firsthand experience. Much of that data collection can be done by *storm spotters*—those who collect storm data from the confines of a laboratory.

While tornado storm chasers are out pounding the backroads of the Midwest in search of that elusive funnel cloud, hurricane experts take a bit of a different tack: they fly into the storms.

Two of the predominant "hurricane hunters" in the United States are the 53rd Weather Reconnaissance Squadron in Biloxi, Mississippi, and the NOAA Aircraft Operations Center in Tampa, Florida. When storms are on the horizon, these crews head into them.

*The 53rd Weather Reconnaissance Squadron in Biloxi, Mississippi, is allotted 10 C-130 aircraft to cover up to five storm missions a day from the mid-Atlantic to Hawaii.*

*(Courtesy of the U.S. Air Force.)*

Hurricane hunters are critical to forecasting the storm's path, strength, and size. While the National Hurricane Center's sophisticated equipment can see the hurricane from space or show by model where the storm is, only by flying into the cyclone can they tell where the pressure center is. The crews also can collect accurate wind speeds within 105 nautical miles of the eye's radius and send that information every 30 seconds.

Flying into a hurricane sounds crazy, right? While the planes aren't modified or strengthened for the task, they are capable of handling pretty tough sustained winds. The hurricane crew doesn't fly over the top of hurricanes—these planes can't fly that high—but it does fly right through them. It's a scary proposition to think of approaching the storm, heading into it, and gunning straight for the eye. The most turbulent the ride gets is just moments before that eye, when the wind speed and strength are at their greatest. Once inside the eye, the crew releases a *dropsonde* into the eye to collect data.

> **def•i•ni•tion**
>
> A **dropsonde** is a small tube with instruments attached by a parachute and sent into the eye of the storm. On its plummet into the sea, it measures the pressure in millibars.

From looking at weather history and studying computer models to chasing down tornadoes and flying through hurricanes, severe storm forecasting can take some pretty extreme routes. Throughout this book, we'll revisit some of the techniques used to forecast and monitor each type of storm. For now, know that flying through hurricanes or chasing down twisters is about the most extreme forecasting regularly practiced.

## The Least You Need to Know

- Early forecasting relied mainly upon watching and documenting weather patterns and learning from observations.
- Modern forecasting uses an array of models, computers, satellites, and other high-tech observations, but still relies on historical weather data.
- Weather forecasting is an inexact art, although much progress has been made.
- Extreme forecasting may mean racing toward a tornado or flying through the eye of a hurricane.

# Chapter 4

# Fronts: How's the Weather Up There?

## In This Chapter

- Putting up a front
- Massive air currents
- Extreme anomalies
- Storms on the horizon

Weather doesn't just happen. It doesn't just rain on a sunny day and tornadoes don't blow in because we're overdue for one. There's a vehicle for all those storms—twisters, blizzards, ice storms, and heat waves—and they are appropriately named "fronts."

They even help us get an idea about what kind of weather is on its way. In this chapter, we'll examine some of those fronts along with a couple more neat weather vehicles and show how they can affect you and me.

# Storm Fronts

Storm fronts aren't always invisible. There are a few good signs we can read, even if we're not paying much attention, that a storm or a severe change in weather is on its way. Wind is a good example. So is looking at the pressure changes on a barometer or the temperature changes on a thermometer. Or even just seeing some nasty black clouds rolling in over the horizon. Yup, fronts. Every extreme storm comes with—well, *in*—one (or more).

Plainly put, there is a boundary that separates two air masses of a different type. That boundary is the *front.*

Lucky for us, there are two major types of major fronts in this world: warm and cold.

## Types of Fronts

A cold front moves into an area that's relatively warmer. When the cold front blows in, the warm air that we were enjoying is pushed up (hot air rises) and it usually gets windy—the wind most always shifts—and we can see cumulonimbus clouds and fast-moving thunderstorms. (See the section "Clouds and You," at the end of this chapter, for a discussion of different cloud types.) A cold front will hustle through an area relatively quickly, moving at speeds of about 20 to 30 miles per hour, then the weather will clear. And it gets chillier, of course.

**def•i•ni•tion**

A **front** is a boundary that separates two air masses of different types.

That summer thunderstorm is a good example. We're enjoying a picnic in the park under clear skies but by the time the bucket of chicken is eaten we notice a slight breeze, look up and see big, fluffy, piled-high clouds on the horizon. It gets breezier and within a half hour we're packing up the kids and fruit salad and making a beeline for the minivan.

On a weather map, cold fronts are drawn as blue triangles pointing in the direction that the front is moving.

*Cold front.*

On the other hand, the warm front is a bit more gradual. It moves in at speeds of about 10 to 20 miles per hour and the warm air lazily pushes into an area occupied by cold air. A few different cloud formations are the giveaways here: cirrus, stratus, and nimbostratus. And with the front comes rain, usually pretty steady and gradual. Weather maps depict warm fronts with red semi-circles linked together, with the round side facing the direction that the front is heading.

*Warm front.*

Cold fronts are the ones that give us those really big windy storms, from thunderstorms to blizzards. Warm fronts provide us with those long summer rains—but park them over an area for a long enough time and the floods come.

Sometimes these fronts get mixed up, and the result is another front that indicates we should put on our galoshes. This is the occluded front. Fast-moving cold air blows into a warm area, overtaking the warm air and quickly pushing it aloft. The result is long periods of rain showers. This one can cause quite a bit of flooding, too, and is usually upon us before we know it. On a weather map, the occluded front looks a bit confused. It's a mix—a pattern of a red semicircle, like a warm front, followed by a blue triangle, like a cold front. The semicircles and triangles both point in the direction where the front is going.

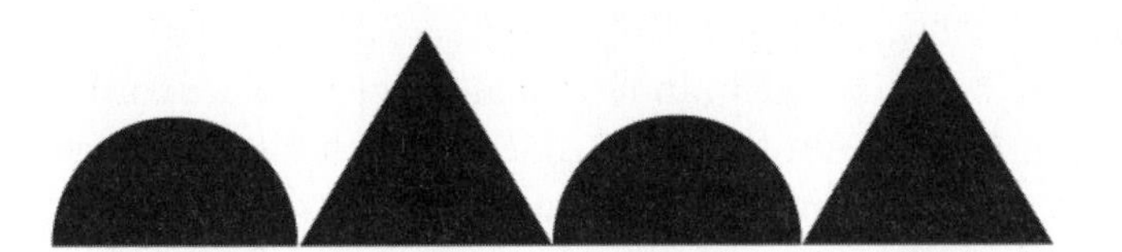

*Occluded front.*

Lastly, the front we really barely notice is the stationary front. Like it's name, it's just a boundary between warm and cold fronts that parks itself over an area and barely budges. There's nothing extreme about this front, but it's worth mentioning. On a weather map, it's just like the occluded front, except the warm red semicircles are facing away from the direction the front is moving and the triangles point toward the area it's moving to, if it's moving at all.

*Stationary front.*

## Fronts and Extreme Weather

The cold front is what we're talking about when we talk about extreme weather fronts. As we learned in the last chapter, air moves from high pressure to low pressure. Looking on a weather map, you'll see the front line, warm or cold, and general areas of high and low pressure. The result of squeezing those two pressure areas is wind. The result of the fronts can give us the rest—the storms. (The fronts are the areas of low pressure on a map.)

It's the atmospheric pressure that drives the weather fronts and storm systems. Cold air is more dense than hot air—the molecules are closer together in cold air, therefore putting more pressure on the ground, for instance, than hot air, whose molecules are more spread out. We can see this over open ocean: the cold air actually depresses the sea level in its vicinity, whereas the warm air elevates it. So that cold, dense air pushes the warmer air upward. When the two collide, that's the front.

Cold fronts are pretty frequent. Think of how many thunderstorms your area sees each year. That's just a small percentage of how many actual cold fronts move through. In fact, at any given moment around the world, nearly 1,800 thunderstorms are occurring. That's about 100,000 a year in the United States alone, and 16 million a year for the entire planet.

def•i•ni•tion

A **severe thunderstorm** must have ¾-inch hail, winds of 58 miles per hour or higher, or be accompanied by a tornado.

The good news is that only about 10 percent are considered *severe thunderstorms*, which means hail, high winds, or tornadoes.

# Jet Stream

A jet stream delivers more than a bumpy ride on that 747; it brings all sorts of weather to areas, depending on its trajectory. Found in the upper levels of the atmosphere, the jet stream is like a river of swift-moving air thousands of miles long and hundreds of miles wide, but pretty thin—just a couple of miles. These streams buzz along about 6 miles to 9 miles above the ground.

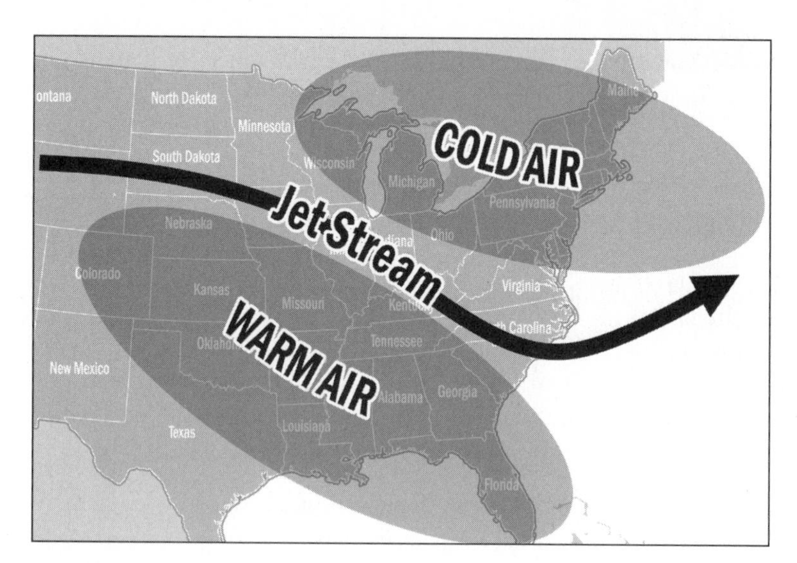


*Jet stream.*

*(Courtesy of the National Weather Service.)*

## Jet Set

*Jet streams* direct storms, basically pushing them along in their paths, and determine high- and low-pressure areas. These conveyor belts are formed between masses of cold and warm air.

When winter temperatures in Montgomery, Alabama, dip into the 20s and 30s overnight, chances are that the jet stream is pushing icy Arctic air down from Canada. The jet stream tends to track farther into the South—even into the Gulf of Mexico—during the winter when it is stronger, and retreats back into Canada most of the summer.

**def•i•ni•tion**

A **jet stream** is a high-altitude ribbon of wind that travels between warm and cold that can drive weather, sometimes severe weather.

This explains much of the deep cold in the Deep South.

Jet streaks, the strongest winds in a jet stream, are indicators that meteorologists seek out. These jet streaks move faster than the winds surrounding them and tend to get the low-pressure areas working into a nice spin as they bolt by in the upper atmosphere. This can lead to some pretty severe storms.

## Raining on Your Parade

We've seen how jet streams work, but how do they feel?

In 2004, the United Kingdom witnessed some miserable beach weather for a good part of the summer. The blame was placed squarely on the shoulders of the jet stream.

Typically, the jet stream that affects European weather passes south of the United Kingdom, leaving it to suffer with mid-60 temperatures and rainy, drizzly weather. If the jet stream moves north, above the U.K., then it brings the warm Azores air that is typical for summer weather. The condition is called an Azores high, and that mid-Atlantic weather is the stuff of great summer days at the beach.

Can the jet stream ever get extreme, though? You bet. In December 2006, in Cape Cod, Massachusetts, a freak weather event made the land look like a tornado had rambled through. Trees knocked over, storm damage, sudden wind—okay, let's back up.

Only a few times in the last 100 years has a renegade gust of wind from the jet stream dove toward earth and knocked down whatever was in its path. That's what happened December 9 on the cape. Scientists say the sudden wind bursts were the product of high-altitude air mass composed of cold air that hit a warmer low-pressure system. The result was the heavy air plummeting to the earth at about 100 miles per hour. In essence, it was the jet stream dead-ending on Cape Cod.

Odd? Indeed.

# El Niño, La Niña

Many people have a good enough grasp of the Spanish language to know that El Niño means "the child." La Niña is the feminine version.

But what's that got to do with extreme weather? Well, these two children can cause major disruptions in weather, with catastrophic results.

It all starts in the western Pacific Ocean, where water cools and warms in cycles. The east-west *trade winds* concentrate the sun-warmed water while the ocean's deep, cold water rises and pushes toward the South American seashore.

**def•i•ni•tion**

**Trade winds** are prevailing winds near the tropics that circulate through the atmosphere, blowing northeasterly in the Northern Hemisphere and southeasterly in the Southern Hemisphere.

Once every few years, however, these trade winds change direction and that warm water moves back east to retard the rise of the colder water. When that happens, weather around the globe is thrown for a loop.

La Niña is sort of the opposite: temperatures become cooler than normal on the ocean's surface near the central and tropical Pacific Ocean. This condition also affects global weather patterns every few years.

**Storm Stats**

El Niño (the warming of the eastern Pacific Ocean) and La Niña (the cooling of the eastern Pacific Ocean) alternate about every three to four years.

El Niño and La Niña actually alternate about every three to four years. La Niña tends to make conditions in the Southwest United States even drier than normal, summer through winter. In the Midwest, the fall sees dry weather, and in the Southeast, it's dry in just the winter. However, areas like the Pacific Northwest become wetter in late fall and early winter, but are colder than normal. In the Southeast, the winters are warmer.

**Eye of the Storm**

El Niño has been blamed for the drought and fires in Australia and Indonesia in 1982–1983 that caused $8 billion in losses to the world economy.

These weather conditions don't make weather severe so much as they deliver it to different areas. You see, these extremes in climate alter jet stream patterns, which can mess with a storm's track and intensity.

In tough El Niño years, monsoon patterns can change, bringing drought and intense heat instead of rain to the Asian countries that depend on precipitation for their survival. Drought brings famine, and thousands of people in India and Bangladesh have starved to death when the wet monsoons brought little rain.

Hurricanes have been blamed on these climate conditions as well, however, in theory. If these conditions can alter fronts, then they may have a play in destructive storms.

So, why the Spanish names for these two little children? The names are derived from their arrivals, which usually come during the Christmas season. Peruvian fishermen dubbed El Niño after the Christ child. La Niña followed El Niño.

# Clouds and You

As we mentioned in the first part of this chapter, clouds are great indicators of what kind of weather is on the horizon. They also cause some weather conditions and result from others.

Here's the rule: when air rises, clouds usually form. The rising air expands and loses heat and as it cools, it can't hold on to as much water as the warmer air can. That water has got to go somewhere, so the little vapor particles band together, form raindrops (or snow), and you get wet if you don't see it coming.

So you want to be a pro at calling the weather? Well, if you know your clouds, you'll at least have a leg up—as well as a larger Latin vocabulary.

**def•i•ni•tion**

The four major cloud groups are **Stratus, Cumulus, Cirrus,** and **Nimbus.**

*Stratus*, *Cumulus*, *Cirrus*, and *Nimbus*. These are the four major types of clouds, and there are a few variations of each of them.

Now for the Latin: Stratus means "layer"; Cumulus means "heap"; Cirrus is "curl of hair"; and Nimbus is "violent rain."

Guess which one is the storm cloud.

The clouds are further broken down (and somewhat combined) into four major categories.

High Clouds are Cirrus, Cirrostratus, and Cirrocumulus; Middle Clouds are Altostratus and Altocumulus; Low Clouds are Stratus, Stratocumulus, and Nimbostratus; and Clouds with Vertical Development are Cumulus and Cumulonimbus.

*Stratiform Altocumulus clouds.*

*(Courtesy of the National Weather Service.)*

Cirrus clouds are long, thin, streaming clouds that usually move west to east. Cirrocumulus clouds are round and small, like cotton balls, and can be grouped or in rows. Cirrostratus clouds are sheetlike and often blanket the sky.

*Cumulus clouds.*

*(Courtesy of the National Weather Service.)*

Altocumulus clouds are gray and puffy. Altostratus clouds also are gray, but cover the whole sky.

Stratus clouds are gray and also cover the sky, and the weather is drizzly. Stratocumulus clouds are lumpy and low. Nimbostratus clouds are very dark gray clouds, almost black, and are low to the ground. Chances are, it's going to rain or snow when you see these clouds on the horizon.

Cumulus clouds look like giant, stacked cotton—flat on the bottom, fluffy on the top.

*Cirrus clouds.*

*(Courtesy of the National Weather Service.)*

Cumulus Congestus clouds are even more puffy than regular old cumulus clouds, looking like the cotton burst open at the top.

Cumulonimbus clouds look like anvils, and when you see these big nasty clouds you can probably hear the thunder in them.

*Cumulonimbus clouds.*

*(Courtesy of the National Weather Service.)*

Knowing your clouds could help you plan your day and impress your friends, but mostly, it will help you know if severe weather is on the way.

## The Least You Need to Know

- Warm and cold fronts are the building blocks of weather and storms.
- The jet stream doesn't just cause weather; it moves it.
- El Niño patterns have been blamed for multi-billion-dollar weather disasters around the globe.
- There are four major cloud groups: Stratus, Cumulus, Cirrus, and Nimbus.

# Part 2

# The Answer Is Blowing in the Wind

No other force on Planet Earth is as devastating as the wind. Registering in the hundreds of miles per hour, tornadoes, hurricanes, typhoons, and sandstorms can wreak havoc so great, whole cities, forests—even bodies of water—can be blown away in just a few minutes.

Unlocking the cyclonic movement of this weather force may give us insight into how to better prepare for storms, as well as ride them out. With all this wind and more people in harm's way, understanding which way the wind blows is critical to protecting life and limb.

## Chapter 5

# Well, Blow Me Down: Hurricanes

### In This Chapter

- You say typhoon, I say hurricane
- Extreme surges
- Some serious storms
- Storm names

Hurricane. Just say the word and some of the adjectives that come to mind—powerful, devastating, destructive, frightening—all have one thing in common: they add up to the most powerful weather force on this planet.

Hurricanes cause billions of dollars in damage, claim scores of lives, and can even wipe entire cities off the map.

Periods of warmer ocean temperatures can bring more severe seasons with more frequent and more powerful storms, too.

In this chapter, we'll look at some extreme hurricanes and their cyclonic counterparts and the havoc these monsters bring with them, including tornadoes, record rain and flooding, and losses in human life and property, as well as why we seem to be pummeled more in recent times.

# What Is a Hurricane?

Simply put, a hurricane is a *cyclone* born in the warm tropics that travels in a low-pressure spin.

A **cyclone** is a large circulation of wind that spins around a low atmospheric pressure region.

It's what that low-pressure system does—how powerful it gets, how long it lasts, and where it strikes—that makes a hurricane memorable.

Hurricanes go by different names all over the globe, and most areas have their version of a storm-producing conveyor belt. These giant cyclonic storms are called "typhoons" in the Northwest Pacific Ocean, west of the dateline; "severe tropical cyclones" in the Southwest Pacific Ocean and Southeast Indian Ocean; "severe cyclonic storms" in the North Indian Ocean; and "tropical cyclones" in the Southwest Indian Ocean.

They all have different seasons to boot, depending on where you live.

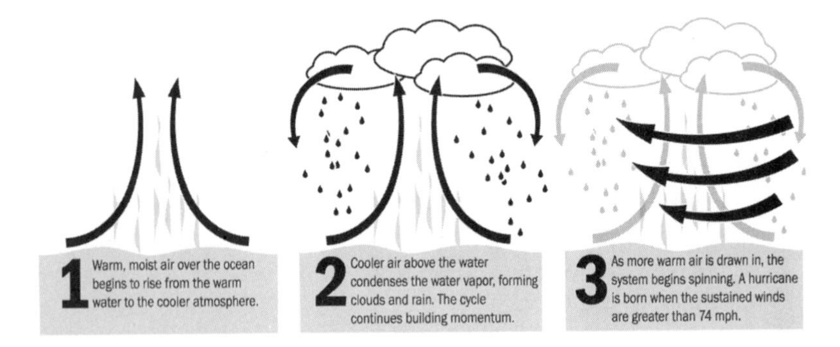


*How a hurricane forms.*

*(Courtesy of the National Oceanic and Atmospheric Administration [NOAA].)*

## How a Hurricane Forms

There is no other single regular producer of hurricanes in North America like the so-called Atlantic conveyor belt. It is exactly what it sounds like: a surge of wind off the northwestern African coast that spirals over increasingly warm water as it heads toward Central and North America, gaining in strength from the warm tropical temperatures. When the winds inside the storm reach 74 miles per hour, it's a Category 1 hurricane. At 96 miles per hour, Category 3, it's considered a major hurricane.

North of the equator, storms spin counter-clockwise; south of it, clockwise. This is due to the Coriolis force, which, if we can revisit our middle school earth science class for a moment, is a product of the planet's rotation. Once air has been set in motion, it becomes deflected from its path as a result of that rotation.

The force and frequency of the storms around the globe vary, but they all are measured on the Saffir-Simpson Hurricane Scale.

**Storm Stats**

Hurricane seasons vary around North America. In the Atlantic, Caribbean, and Gulf of Mexico, it's from June through November; in the eastern Pacific, June through November 15; and in the central Pacific, June through October.

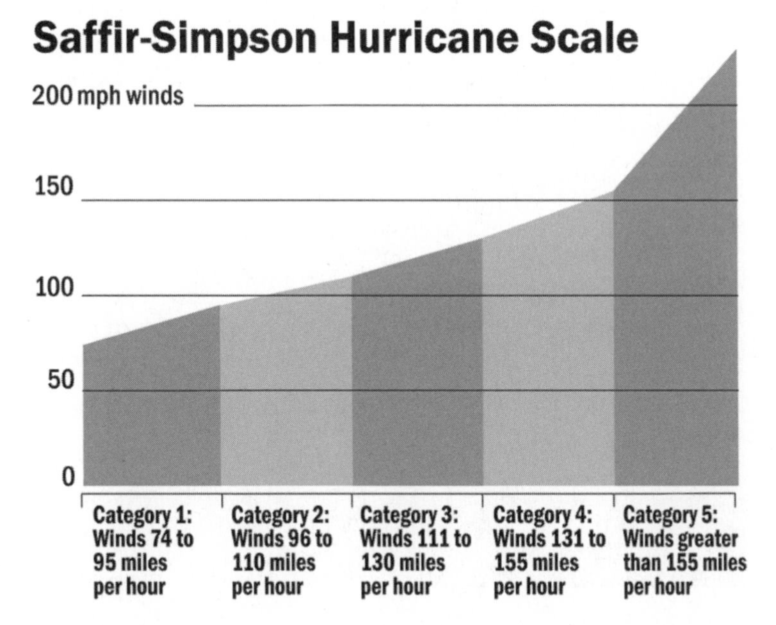


*Hurricane categories.*

*(Courtesy of the National Oceanic and Atmospheric Administration [NOAA].)*

So why not a Category 6, 7, or 12? First of all, storms are extremely hard to measure even in the lower Category 5 range. Simply put, the equipment to measure the storms becomes damaged or is blown away. For years, experts claimed the Northwest Pacific's Typhoon Nancy in 1961 reached maximum sustained winds of 213 miles per hour with a central pressure of 888 millibars. These were estimates based on a model that has become somewhat obsolete. Scientists today agree that the data collected from the 1940s to 1960s were a bit on the high side. Data compiled since put atop the list storms like Hurricane Camille in 1969 and Hurricane Allen in 1980 with winds at 190 miles per hour. Again, wind measurements are iffy when wind speed is at the point where it's beating up equipment, so estimates still play a role.

## Storm Strength

As a hurricane chugs across the warm Atlantic, it picks up energy. Forecasters keep a keen eye out for these storms and issue weather bulletins: *hurricane watches* and *warnings*. The warmer the surface water, the stronger it can make the storm. When the hurricane strikes land, like the islands of the Caribbean—often the first barriers before continental strikes—they tend to lose steam, but not always. Ditto for cooler waters. When a storm tracks north where it finds water below about 80 degrees Fahrenheit, it generally weakens. And when the storm does hit the continent, its size, strength, and composition diminish rapidly compared with the life of the storm over sea, again because there is no warm-water power source.

**def•i•ni•tion**

A **hurricane watch** indicates that hurricane conditions could occur in an area within 36 hours. A **hurricane warning** indicates that winds of at least 74 miles per hour are expected within 24 hours.

*A satellite image of Hurricane Katrina, still at Category 5 strength, before landfall on the Gulf Coast.*

*(Courtesy of the National Oceanic and Atmospheric Administration [NOAA].)*

It is rare for a hurricane to remain as such just a few hours after it strikes landfall. Oh, but the coastal damage from that wind ….

Hurricanes that skirt through the Caribbean or smack the eastern Florida coast head-on can be brutal. But they don't have to be major (Category 3 or higher) storms. Even a Category 1 storm that strikes a particularly sensitive area directly can be devastating—knocking down trees, cutting power, causing flooding, and even claiming lives. Folks don't always take seriously enough the force of even a smaller hurricane, choosing to "ride it out" or not bothering to prepare their property. Category 2 Floyd in 1999 claimed 56 lives in southern Florida despite hurricane warnings and evacuation orders. Hurricane Agnes in 1972 killed 122 people in Florida and, later, in the Northeast. Agnes, however, was a mere Category 1 storm, but still ranks twenty-third of the most deadly American hurricanes.

Hurricane Andrew, a Category 5 beast that set the record in 1992 for the most costly cyclonic storm to hit the United States, caused $21 billion in damage when it pounded southern Florida. Damages from Hurricane Katrina in August 2005 are still being tallied, but have topped estimates of $125 billion.

**Storm Stats**

Why not nuke 'em, as suggested by urban legend? Besides that trade winds would carry deadly radioactive matter to entire continents, the power needed to even match a hurricane would be immense. The heat release of a well-formed hurricane is equal to about a 10-megaton nuclear bomb exploding every 20 minutes.

## Storm Duration

It's not just how strong the storm is, it's how fast or—more to the point—how slow it moves. When a storm is tracking toward the coast, what you don't want is it moving so slow that it will park itself over a city for hours on end. The flooding from intense periods of rain and the storm's related surge (more on that in a moment) make what could have been a more minor hurricane in strength now a catastrophic flood.

There's a little island in the South Indian Ocean named La Reunion Island that is home to some pretty big cyclone-related rainfall records. In a 24-hour period, Tropical Cyclone Denise dropped 71.8 inches of rain in January 1966. For 48 hours in April 1956, the little island saw 97 inches from an unnamed tropical cyclone. More recently, 10 days of rain associated with Tropical Cyclone Hyacinthe in January 1980 dropped 223.5 inches of rain.

And you thought Seattle was rainy ….

Other cyclonic byproducts include tornadoes. As if hurricanes aren't bad enough, the spin cycle of these storms tends to manufacture severe conditions over land as their bands, the broad sweeping tentacles surrounding the *eye* of a hurricane, reach out and disrupt the air around it.

It is very common for tornado watches and warnings to be posted when a hurricane makes landfall.

**Storm Stats**

The eye of a cyclone is the center of the storm and often has the lightest winds within it. Conversely, the eyewall is just outside of the eye, and has the greatest force of wind, closest to the storm's axis.

From September 17 through 20, 2004, Hurricane Ivan spawned 117 tornadoes. That's the record for the most tropical storm–related tornadoes. In Virginia, 37 twisters were counted, along with 25 in Georgia, 18 in Florida, 9 in Pennsylvania, 8 in Alabama, 7 in South Carolina, and 4 in both North Carolina and Maryland. When the three stormy days were over, 8 people were dead and 17 injured from the tornadoes alone.

## Storm Surges

Have you ever seen a dolphin on a mountain top? Of course not. But residents of Bathurst Bay in Australia reportedly did in 1899. The Bathurst Bay Hurricane, also known as Tropical Cyclone Mahina, produced a 42-foot *storm surge* that pushed marine life atop some area cliffs.

**def•i•ni•tion**

A **storm surge** is the rush of the ocean caused by the high winds of a cyclone, while a **tidal surge** is a result of the normal ebb and flow of the ocean.

Just for a moment, visualize how tall 42 feet is. Stand next to a four-story parking garage. Now imagine it completely submerged.

So what produces a storm surge, and how come it's the number-one killer of humans during hurricanes?

A storm surge, as opposed to a regular *tidal surge*, happens when seawater associated with storm winds is blown ashore. While storm wind damage and even tornadoes can be major punishers, the storm surge is the Grim Reaper. Nine out of every 10 hurricane deaths, historically, come at the hands of a storm surge.

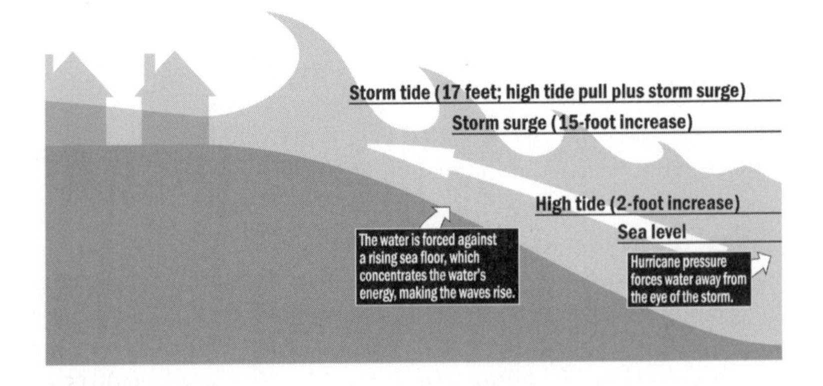


*Storm surge.*

*(Courtesy of the National Oceanic and Atmospheric Administration [NOAA].)*

Living on the U.S. coast has some great advantages, but considering where most hurricanes strike—Gulf and Southeast coasts—and that their coastlines are at an average of 10 feet above sea level, risk from storm tides and surges are great.

Forget that Hurricane Katrina struck the low-lying New Orleans area downgraded from a 160 mile per hour Category 5 hurricane. As it made landfall August 29, 2005, at Category 4 strength with 145 mile per hour winds, it still carried with it the kind of storm surge associated with a Category 5.

The storm surge closest to the storm's eye, Bay St. Louis, Mississippi, was measured at 11.45 feet.

But in metropolitan New Orleans, Lake Pontchartrain, just a foot above sea level, was overwhelmed by the powerful surge and rose to 8.6 feet above sea level. On the following day, the city's levees breached and 80 percent of the city was submerged.

Don't forget that water is heavy—1,700 lbs. per cubic yard. So when the waves of a storm surge come crashing onto the beach, practically anything in their way won't be there for long. Beach erosion also is common during a storm—and not just for those areas that take direct hits.

When Hurricanes Ophelia and Tammy swept by the South Carolina coast in September 2005, they each took with them 5 feet of sand from the Hunting Island State Park beaches just north of Hilton Head Island. In a normal year, the historic beaches lose 15 feet of sand, but hurricanes, as we've seen, can greatly compound that number.

**Storm Stats**

If the winds or waves related to a hurricane don't tear down beach-front buildings, eroding sand from a surge can put them on faulty ground, deeming them uninhabitable.

# Extreme Storms

Two weeks after Hurricane Katrina struck, much of the Gulf Coast was the bull's-eye for another Category 4 storm: Hurricane Rita. Fortunately, the area already had been evacuated from Katrina, and casualties were far less.

Two extreme hurricanes within a month. Freak storms, right? Maybe not—at least not according to weather experts. In 2004, one in every five homes in Florida was damaged by a hurricane. A more active Atlantic conveyor belt produced one of the busiest hurricane seasons in Atlantic basin history, with 15 tropical storms and 9 hurricanes, including 6 major hurricanes that forced the evacuation of tens of thousands, caused an estimated $42 billion in damage, and claimed the lives of 117 in Florida alone. However, flooding from the storm surge of Category 3 Jeanne in Haiti left more than 3,000 people dead.

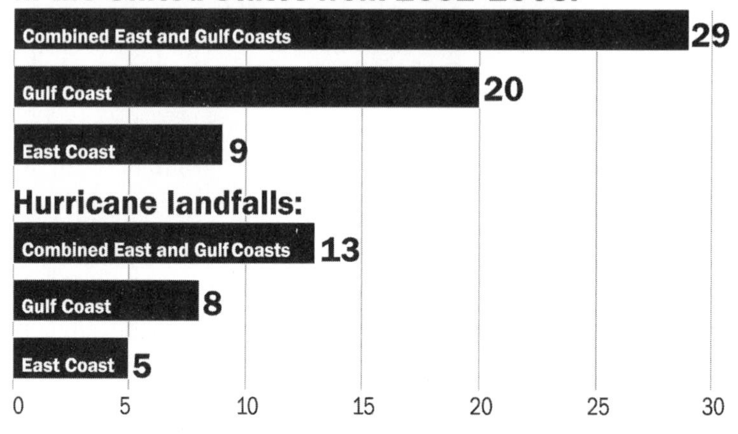


*Hurricane strikes.*

*(Courtesy of the National Hurricane Center.)*

From 1944 to 2004, the average number of storms per year in the Atlantic basin—the Caribbean to the lower Southeast coast of the United States—included approximately 10 named storms and 6 hurricanes, with 2 to 3 major hurricanes.

**Eye of the Storm**

Camille holds the record for strongest winds, with sustained winds of 190 miles per hour. Katrina held tight in Grand Isle, Louisiana, with sustained speeds at 140 miles per hour.

Only two hurricanes with stronger sustained winds than Katrina have made landfall in the United States: The Labor Day Hurricane of September 2, 1935, was a Category 5 nightmare that struck the Florida Keys at 892 millibars; Hurricane Camille, on August 17, 1969, brought its wrath upon Mississippi as a Category 5 storm with a 909 millibar reading. Southeast Florida's Andrew in August 1992 was at 922 millibars, just 2 millibars weaker than Katrina's 920 millibars.

Only four U.S. storms have claimed more lives than Katrina's estimated 1,300. Hurricane Galveston in Texas remains number one. The unexpected Category 4 cyclone of 1900 claimed an estimated 8,000 lives. The Florida Hurricane of 1928 that struck the southeastern Florida coast and into Lake Okeechobee killed an estimated 2,500 people. Louisiana's Category 4 Cheniere Caminanda Hurricane in 1893 killed an estimated 1,100 to 1,400 people. Although estimates are sketchy on the Sea Islands

Hurricane of South Carolina and Georgia, also in 1893, the Category 3 storm killed an estimated 1,000 to 2,000 people.

In comparison, a Category 3 hurricane in Puerto Rico in 1899 killed 3,369 people and, as we've mentioned earlier, the Bangladesh Cyclone of 1970 killed an estimated 300,000 to 500,000 people, most of the victims having drowned from the powerful storm surge.

Many poorer or Third World nations, with massive coastal populations, unfortunately, are home to some of the greatest casualty numbers on the globe, as was the case in Bangladesh in 1970 and again in 1991 when 138,000 perished in cyclonic winds of 138 miles per hour and a 20-foot storm surge.

## What's in a Name?

It's a funny thing, this naming of storms. We don't name tornadoes, blizzards, or mudslides. It would seem silly. But we do name our hurricanes. And on the opposite corners of the stormy planet, folks name their cyclones, too (although with sometimes more meaningful or symbolic names).

The naming of hurricanes began during World War II when Army Air Corpsmen and Navy meteorologists named the storms after their girlfriends or wives. From 1950 to 1952, the U.S. government began naming the storms phonetically for clarity. A year later, the U.S. Weather Bureau began using female names, and in 1979 the National Weather Service began including male names.

Tropical cyclones in the Northeast Pacific also began being named in 1950, and in the Hawaiian islands, a year later. It wasn't until 1979 that they took on male names, too. Over in the Northwest Pacific, tropical cyclones were named after women in 1945 and after men, too, in 1979.

As of the beginning of the twenty-first century, those tropical cyclones began being named with Asian monikers that were contributed by all the nations and territories that are members of the WMO Typhoon Committee. The names may be after animals, like a lizard, or a tree, for instance. Confusing to some, the names are not in alphabetical order by name but by contributing country.

Beats Hurricane Hazel, right?

As for cyclones in the North Indian Ocean, there are no names, but the cyclones in the Southwest Indian Ocean have been getting names since 1960.

The Aussies named their storms after women since 1964 and, progressively, began including men in 1974.

So back to ol' Hazel for a moment. Several hurricane names have been retired. But not because there aren't a whole lot of Hazels or Agneses being christened anymore; it's that these were extremely violent storms, and to differentiate between the deadly Hilda of 1969 and another, less intense Hilda, the names were put to pasture to avoid confusion.

> **Storm Stats**
>
> You won't find hurricane names that begin with Q, U, X, Y, or Z because there aren't many names that begin with those letters. No Quasar, Xavier, Yolanda, or Zelda.

For Atlantic Basin storms, there are 21 names chosen each year.

The list runs six years out and, until 2005, a year's worth of names never had been exhausted. When that does happen, then the Greek alphabet is used. Epsilon was the last of 2005, and was born and died with little fanfare over the Atlantic in December.

You won't find far-out names: a prerequisite for storm names is that they are short and easy to pronounce by broadcasters—at least in the United States. So, while folks may have stumbled on Hortense or Humberto, you won't get any tongue-twisters.

So Hurricane Bob might be boring, but it's functional. And besides, if American names are not as cool as the Asian ones, consider this: rather than just having named the heavy weather force "cyclone," named for the storm's spinning characteristics, Americans looked for a name strong in legend.

"Hurricane" is derived from the Carib god of evil. His name? Hurican. Speaks volumes, doesn't it? The Caribs, however, ripped off the name from the Mayans, who have Hurakan, a creator who, legend has it, blew his breath across the water and made the land dry. He also later destroyed men with storms and floods.

Go figure.

"Typhoon," you ask? It means "big wind" in Chinese. Fair enough.

## The Least You Need to Know

- Hurricanes form when warm tropical water is drawn into a spinning vortex, generating a stronger cyclonic low-pressure movement.
- Storm surge is the number-one killer of people during a cyclone.
- Extreme cyclones are characterized by their size, strength, speed, duration, and how much havoc they wreak on humans, property, and the environment.
- Cyclonic storms are named to avoid confusion.

# Chapter 6

# Spin Cycle: Tornadoes

## In This Chapter

- Tornado formations
- Extreme twisters
- Tornado country
- Twisters around the world
- Riding out the storm

Tornadoes, like other cyclonic storms, can be unpredictable. They can grow in strength and size in the blink of an eye and change direction just as quickly. For that reason, they always must be taken as a serious and deadly threat. The lay of the land, climate, and parent storms play major roles in the generation of these turbulent weather events. In this chapter, we'll take a look at how, where, and why severe tornadoes form, and just how damaging they can be.

## How a Tornado Forms

To this day, thousands, if not millions, of people think of that big, black, evil-spinning twister of *The Wizard of Oz* fame when they hear the word *tornado*.

They aren't all that far off.

Although cows making eye contact with you as they zip around the vortex and green-faced wicked witches riding bikes or brooms is far-fetched, the same sort of menacing twister that slowly bounded toward Dorothy's Kansas farmhouse home could easily—and often they do—pick up entire houses, not to mention cars, trucks, trees and, yes, even cows.

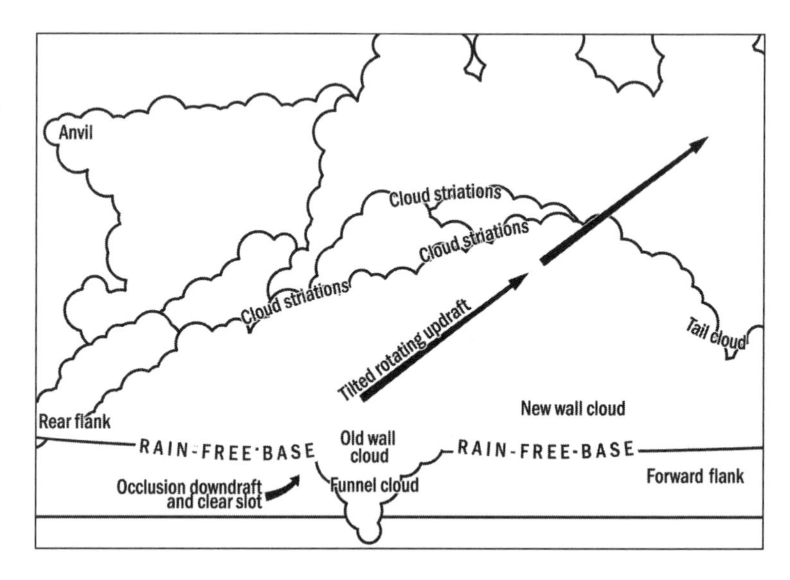


*Supercell formation.*

*(Courtesy of the National Severe Storms Laboratory.)*

But unlike the fictitious tornado that transports Dorothy to Oz, few people ever live to tell about being caught in the storm's path. The violent force of a tornado can take down the most solid of dwellings and blaze a swath through forests that looks like the department of transportation was cutting a new road with bulldozers.

**def•i•ni•tion**

A **cumuliform** cloud has a flat base and rounded outlines, piled up like a mountain.

Whereas hurricanes are predicted days or even weeks ahead of their landfall, tornadoes can form suddenly in a rotating thunderstorm, known as a supercell thunderstorm.

By definition, a tornado is a rotating column of air attached to or beneath a *cumuliform* cloud. That column must be in contact with both the cloud and the earth if it's going to be called a tornado.

Of course, these vortexes, whether they touch the ground, can be devastating. It's not that uncommon, either, that one pillar forms into another or splits off, forming what is called a multiple-vortex tornado. And these twisters can be spindly, whipping like a wispy rope, or immense in girth—even as much as one or two miles wide!

*A mature tornado is seen in Enid, Oklahoma, on June 5, 1966. Note the connection to both the ground and cumuliform clouds.*

*(Courtesy of the National Oceanic and Atmospheric Administration [NOAA].)*

*Aerial view of the Moore-Oklahoma City tornado path on May 5, 1999. Note how the tornado swerved to cross the highway before moving left again.*

*(Courtesy of the National Weather Service.)*

Okay, so now we all know what a tornado is, but there is a bit of mystery about how these bad boys form. Sure, warm, wet air meets cold, dry air during an *occlusion downdraft* is the age-old recipe, but those ingredients also make run-of-the-mill thunderstorms, and not every thunder-boomer spawns a twister. Temperature, rotation, and downdraft are common to tornadoes, but how and if temperature variation actually plays a role, or how much rotation in thunderstorm clouds are, in fact, tornadic, is the mystery.

**def•i•ni•tion**

**Occlusion downdraft** is an arc of sinking air believed to contribute to tornado development.

Those rotating mesocyclone thunderstorms—so-called supercells—are the most severe of all thunderstorms. They contain hail and lightning and can cause flash floods.

Although Midwestern thunderstorms are the most common producers of tornadoes, tropical storm systems can play a role in hurricane proliferation. Hurricane Katrina in 2005 and Floyd in 1992 spawned several tornadoes throughout the Deep South. Again, that warm, wet tropical air mixing with the relatively cooler, drier air is ripe tornado weather.

The most notorious of tornado outbreaks in recorded history, however, spanned from central Kansas, through eastern Oklahoma, and on into Texas on May 3, 1999. A staggering 70 tornadoes touched down, killing 40 people in Oklahoma, the most heavily hit area, and caused $1.2 billion in damages. The event has become known as the Central Oklahoma Tornado Outbreak of 1999.

Nearly 2,000 houses were destroyed and about 6,500 were damaged. That doesn't include apartment, business, or public buildings. The series of twisters reached enormous storm strength as well, with measurements logged at F5 on the Fujita Scale of Tornado Intensity.

Winds inside an F5 tornado—the maximum measurement on the Fujita Scale—can reach a dizzying 318 miles per hour.

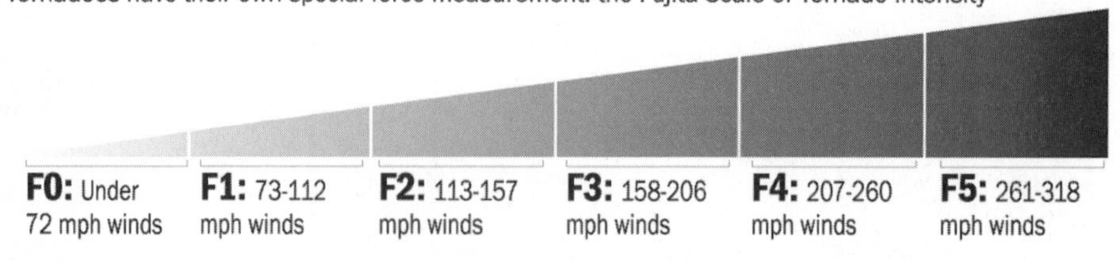


*Although the 70-twister outbreak was the largest, it wasn't the deadliest. Several tornadoes throughout history and around the world have claimed hundreds of more lives.*

*(Courtesy of the National Weather Service.)*

## Most Damaging Tornadoes

Although there are several ways to measure a tornado's wrath, the cost of human life outweighs monetary and property damage and storm strength every time.

For just under four afternoon hours on March 18, 1925, residents of southeastern Missouri, southern Illinois, and southwestern Indiana knew they were in trouble when the rain preceded a powerful thunderstorm and the skies blackened.

A farmer was killed when a tornado touched down in a rural area just northwest of Ellington, Missouri. His death was the first of 695 deaths along the 219-mile continuous track of the Great Tri-State Tornado.

Traveling at speeds between 62 and 73 miles per hour, the F5 tornado's wind speeds reached in excess of 300 miles per hour and grew to widths of a mile wide, according to historical accounts. The barometric pressure measured 28.87 inches.

The storm lifted houses off their foundations, razed barns, and killed livestock. In the end, 234 people died in Murphysboro, Illinois, which stands as the record for tornado deaths in a single community. A school also was in the path, and 33 people, mostly children, perished. An estimated 15,000 homes were destroyed and 2,027 people were injured in the storm.

**Eye of the Storm**

The mile-wide swath of the Tri-State Tornado leveled an entire town. Gorham, Illinois, was demolished and 34 residents died in a matter of moments.

The Tri-State Tornado finally weakened and dissipated just north of Princeton, Indiana, more than 200 miles and three hours from where it began. Damage to property totaled $16.5 million.

The storm stands today as the most deadly, and does so by a long shot. In 1840, a tornado in Natchez, Mississippi, claimed the lives of 317 people. In St. Louis, Missouri, 255 people perished in 1896. A 1936 tornado in Tupelo, Mississippi, took the lives of 216 people, and 203 Gainesville, Georgia, residents were killed in a 1936 twister.

All of the top 25 most deadly American tornadoes happened before 1956, most occurring closer to the turn of the twentieth century. Tornado warning systems and technology since the mid-1950s have made great strides in protecting people from these vicious storms.

**Storm Stats**

How strong can a tornado get? Typical tornado monitoring devices in the path of a twister can measure up to about an F5 storm. Anything over that and the device is damaged by the storm's powerful winds.

# Tornado Alley

Murphysboro, Illinois; Natchez, Mississippi; St. Louis, Missouri; Woodward, Oklahoma; Purvis, Mississippi; New Richmond, Wisconsin; Flint, Michigan … most of the deadliest tornadoes strike from northern Texas into the Midwest.

The flat terrain unobstructed by mountains and the climate where warm, wet air meets dry, cold air is as close to perfect for tornado generation as can be. For this fact, most of the western Midwest into northern Texas has been dubbed Tornado Alley.

*Tornado Alley.*

*(Courtesy of the National Weather Service.)*

But the three letters that are most synonymous with tornadoes are OKC—Oklahoma City, the capital of Tornado Alley. Since tornadoes have been recorded, experts agree that this city in the heart of the nation's breadbasket stands as the national bull's-eye for twisters. Between 1893 and 2004, a staggering 116 tornadoes have struck OKC, ranging from rope-like skippers to the massive F5 storm of May 3, 1999. In fact, since 1950, there have been only four periods of more than two years that a tornado hasn't struck Oklahoma City.

On March 28, 1948, an F3 tornado ripped through Will Rogers Field at Tinker Air Force Base, destroying 54 aircraft and causing $10.25 million in damages.

Another tornado, on April 30, 1949, claimed 48 lives when the slow-moving twister moved through the North Campus of the University of Oklahoma, destroying well-constructed buildings in its 22-mile raging path.

It's important to note here that tornadoes happen all over the United States—just not as frequently as in the central part of the country. The twentieth most deadly tornado on American soil was in Worcester, Massachusetts—not exactly in the breadbasket. On June 9, 1953, 90 people were killed when a twister ripped through this New England town.

Other odd areas where tornadoes have sliced through towns and claimed lives have been in Florida, almost every state east of the Rockies, and in high elevations like the Appalachians, Rockies, and Sierra Nevada.

On average, tornadoes kill 60 Americans each year. Falling (or flying) debris is the main culprit. But just as tornadoes strike all over the United States, they also occur around the world—but more on that in a bit.

**Storm Stats**

The site of the highest elevation tornado on record, which happened in the Sequoia National Park on July 7, 2004, was at a staggering 12,000 feet!

Typically, if there is a tornado season, then late winter to midsummer would be it. Tornadoes can happen at any time of the year, should the conditions warrant, but June and July are the months when tornadoes strike most frequently in the Midwest. In the southern plains, however, May and early June are the most frequent times. And on the Gulf Coast, early spring is most seasonal.

Because thunderstorms can move in all directions, tornadoes can appear from any direction. Following the cyclonic rules that apply to hurricanes, tornadoes also spin counterclockwise north of the equator and clockwise south of it. Most tornadoes, however, track from the southwest to the northeast or west to east.

## Extreme Tornadoes Around the World

Tornadoes aren't native to only the United States or North America. Anywhere that conditions are favorable, a tornado—even an extreme tornado—could occur. High elevations, low elevations, coastal, interior, even Arctic areas have been witness to tornadoes.

**def•i•ni•tion**

A **tornado watch** means that severe weather is possible and people should be prepared and alert. A **tornado warning** means that a tornado has been spotted (or it looks like one has formed on radar), and, if you're in the warning area, take cover.

Although the United States is home to the most tornadoes on the planet, a tornado can touch down almost anywhere. However, most other nations do not have a specific *tornado watch* or *warning* system in place.

Canada does have such a system, although the areas most affected by the storms are largely unpopulated—the Canadian plains, for instance. Tornadoes may rumble through the central region, with no one ever witnessing an account.

The Meteorological Service of Canada does handle storms and warnings in the northern nation. Such was the case when an F4 roared through Edmonton, Alberta, on July 31, 1987, and another in Barrie, Ontario, on May 31, 1985. Although twisters in Canada generally aren't as strong as their U.S. cousins, these two examples were big storms, to say the least!

Tornadoes can also occur halfway around the world. On March 3, 2005, residents of Greymouth, New Zealand, witnessed a major tornado that tore through their town. According to reports, the twister tracked right through Greymouth's business district at 1 P.M. and destroyed or damaged 48 buildings. Three people were injured and no one was killed, but for residents not used to seeing a giant tornado, the weather event was startling.

**Storm Stats**

Despite modern technology, the number of tornadoes occurring around the planet in a single year is unknown. Even in the United States, there are only estimates. Tornadoes average 1,000 a year in the United States.

Just south of Hong Kong International Airport, a tornado tore through sections of Hong Kong on September 6, 2004. Many at the airport, including travelers and workers, witnessed the account. No one was hurt, but the rare twister was certainly a spectacle.

These two foreign accounts show relatively weak twisters, in comparison to the nasty F4s and F5s of the American Great Plains, but not every tornado in other parts of the world is timid.

An estimated 70 people died and 2,000 were believed to be injured in the early evening of April 14, 2004, in Netronkona, Bangladesh, when a violent storm raged through more than 20 villages, scattering debris and bodies. The poorly crafted, modest homes and buildings were no match for the force of the storm, and officials said tens of thousands of people were left homeless. The storm's winds were estimated at 150 miles per hour.

Even tropical islands can bear witness to tornadoes. In Kunia, Hawaii, in the Central Oahu area, tornadoes touched down on January 25, 2004. Though the 2 P.M. storm caused no injuries or deaths, heavy rain and flooding, lightning, and inch-sized hail plagued the region.

# Hunkering Down

The National Weather Service is the only public tornado forecaster in the United States. When a storm is near, NWS will issue a watch or warning, usually to local media emergency departments, and/or by police-band scanner, computer, or fax. The local media outlets, such as TV and radio, can promptly issue a weather bulletin that will give folks a running head start. Emergency dispatchers also will send out a bulletin over the scanner and, in some cases, TV and radio, and would get the word out to police on the street.

Some local governments use tornado warning sirens, while others do not. There is no national law requiring it. And while the idea of having a computer or phone-based warning system in place has some merit, logistically, it might not be possible. When tornadoes hit, they tend to knock out power and phone lines—both of which are necessary to deliver that information through those mediums.

So when a potential tornado is on the horizon, get the heck out of the way! Most victims of tornadoes are injured or killed by falling debris, so it's important to take cover when a tornado warning has been issued.

## Take Shelter

The stronger the structure, the better off you're going to be in riding out the storm. Many Midwesterners have tornado bunkers or storm cellars in their homes or on their properties for this reason. So-called safe rooms—reinforced small interior rooms that are steel- or concrete-reinforced—offer even more protection. If no safe room, bunker, or basement exists, interior bathrooms or closets might offer protection. The idea is to be deep inside a house with many beams and supports surrounding you—and away from windows.

And opening those windows to "equalize the pressure" is just silly. Should a tornado be that close, it's going to take the windows—and probably the house—with it. Another falsity is which corner of a basement a person should hunker down in to best beat the storm. The best corner to be in is the one without the stove or entertainment center directly above you. Tornadoes can strike from any direction, not just the southwest, and when a tornado blows through a house, it's going to whip heavy furniture around like leaves on a blustery day. Best to get under a sturdy structure, like a workbench, or tuck under a mattress, and be sure granny's armoire isn't overhead.

Unfortunately, we're not always in the safety of our homes or near a buddy's bunker when a tornado strikes. Quite possibly, we may be in a mobile home, which are among the least safe places to be in a tornado. In fact, most tornado deaths occur in mobile homes.

When your home or office isn't a solid structure, then seek out a community shelter in your area. Municipal governments should be able to provide a list.

## Roadside Assistance

If you're caught on the road when a tornado is bearing down, find a safe shelter or drive out of the storm's path if you have time. Bridges make horrible shelters in a tornado. They can collapse, peel away, or even break apart, causing dangerous objects to whip around. Also, they cause great wind tunnels that can suck people clear out from under the bridge.

Cars are complete deathtraps if the tornado is upon you. No bridges, no cars. Vehicles, no matter what their weight, can be tossed about or sucked up by a tornado with little effort. If there's no safe building to get to and no way to drive out of the path of the storm, your chances for survival are better getting as far away from your car—and others—as they can become flying objects, and lying flat in a low spot.

**Inside the Storm**

Compare the tornado to a fixed object to determine its direction, then head away at a right angle. If it's growing larger or appears to be staying in the same place, it's heading right toward you.

In open country, though, if the tornado is off in the distance, try to drive away from it.

The key thought here is getting as much distance between you and that twister.

## The Least You Need to Know

- Tornadoes usually form when cool, dry air meets warm, wet air.
- Loss of lives, not size of storm, equals the worst tornado.
- Oklahoma City is the capital of "tornado alley."
- Tornadoes occur all over the planet.
- Get out of the tornado's way!

Chapter 7

# Waterspouts

## In This Chapter

- Going for a spin
- Transfer of power
- Fish tales
- Don't rock the boat

If you saw someone standing in a Kansas cornfield watching a tornado from a half-mile away, you'd think they were nuts for not running for cover.

But see a waterspout a half-mile over the Gulf of Mexico, and it's "Honey, go grab the camera!"

Why is that?

Well, not too many people are killed by waterspouts. But there have been a few injuries and even damage to boats and homes.

Let's take a look at waterspouts and see what the extremes are with this weather occurrence.

# How a Waterspout Forms

A tornado over water? That's exactly what a *waterspout* is. It's just that a waterspout is a pretty weak form of a tornado—a nonsupercell, to be precise.

**def•i•ni•tion**

A **waterspout** is a weak tornado that spins over water.

But to be even more precise, you can't call them tornadoes unless they hit terra firma—good, solid ground.

Over open water, these twisters can damage ships and flip small boats, and even kill people. And, if they hit land, they can turn into tornadoes. So why do we gawk?

Well, there's a belief that these things don't actually hurt or will just stop when they hit dry land. False.

If you're in your motorboat looking for that 5-pound bass when a twister is bearing down, you're in for a rough, rough ride. Put down the rod and head for shore.

## Waterspout Formation

Despite that waterspouts are tornadoes over water, the way they form and move is a bit different. The air inside a tornado and waterspout moves the same way—upward and counterclockwise, in most cases.

The air inside a waterspout is going pretty fast, though. Wind speeds of up to 190 miles per hour have been recorded.

A waterspout begins in a unique way. A dark spot forms on the water's surface and begins spinning upward while the wind shifts and increases. The warmer the water, the better change for funnel-cloud proliferation. The same goes for the air temperature: the waterspout's formation depends also on cooler, moist air high above the water's surface. A funnel cloud above will begin to drop down in a faint vortex, looking to meet the rising water from the surface. Once its speed reaches about 40 miles per hour, that wind starts to generate spray. The more humid the climate, the better the possibility for funnel clouds to appear over water, since there are more water droplets in the moist air to rise into the cooler clouds.

**Eye of the Storm**

A giant waterspout in the Florida Keys in 1969 had a diameter of approximately 90 feet. Big for a waterspout!

*A waterspout makes its way across Niles Channel near Big Pine Key in Florida on August 20, 1999.*

*(Courtesy of the National Weather Service.)*

Waterspouts can last up to about 20 minutes or as little as 2 minutes. They move at speeds of about 10 to 15 knots.

Once the cloud meets the water via a funnel, the waterspout is in what's called the mature stage. Soon, the vortex grows weak and the waterspout begins to dissipate when the cool air from the eventual falling rain cuts off the supply of warm, humid air.

## Predicting Waterspouts

Waterspouts tend to generate just as we're putting on that second coat of sunscreen. The hottest part of the day, just past noon, is ripe for spouts because the surface temperatures are hottest during this time. When the temperatures are warmer, they cool off faster as they rise—remember, hot air rises.

*Waterspout formation.*

*(Courtesy of the National Weather Service.)*

Can meteorologists forecast waterspouts? As we've mentioned, the conditions for a waterspout are pretty common every day in southern Florida. But some conditions that meteorologists look for when forecasting waterspouts are warm water temperatures with cooler air aloft. The air above is also moist. Wind speeds are usually light.

# Fish in the Wal-Mart Parking Lot

Once in a great while, you hear the stories. Newspapers have them; people talk about them from behind lunch counters over egg-salad sandwiches.

Strange things fall from the sky. Like fish.

> **Inside the Storm**
>
> Fish, frogs, bugs, and birds have been reported to have mysteriously fallen from the sky. Chances are that a waterspout or cyclone took them for a spin.

When waterspouts skip across water, they sometimes can whip up fish and other small sea life, much like a tornado can pluck a goat from a pen, or an oak tree from its roots, and send them aloft to some great distance away.

One thing to remember, however, is that any vortex tends to throw things sideways and not necessarily "suck" them up like a vacuum into the heavens.

There are accounts of people, even, being thrown a great distance in a tornado. One man was found a mile from where he was snatched up by a twister. It happened in 1930 during a hurricane in May. Unfortunately, he died shortly after landing.

Animals certainly have been picked up by tornadoes—ducks found 80 miles from where they were plucked, even cows sent reeling. Maybe this is where we get the phrase "It's raining cats and dogs …."

But waterspouts are much weaker. Whether they would pluck a man from a surfboard is probably pushing it, but you could bet it would be a rough blow into the water, and the force and wind and any other debris caught in that vortex could be deadly at such high speeds.

> **Inside the Storm**
>
> Cyclones certainly can suck up objects, but more than likely, they push them sideways. The spinning force, like that of a washing machine, tends to push things away from it.

But it stands to reason that if smaller fish were caught in that vortex and remained light enough in the water and vapor, when the waterspout hit land and dissipated, those fish would be launched back to earth, and possibly found scattered in the Wal-Mart parking lot.

*A waterspout in its mature stage whirls near Big Pine Key, Florida, in 1999.*

*(Courtesy of the National Weather Service.)*

Accounts have it that in 1939 people in Wiltshire, England, were pummeled from the sky by small frogs.

Certainly not out of the question.

Other odd accounts include snails falling from the skies above Chester, Pennsylvania, in 1870. The same thing happened in Nigeria in 1953. Ducks and woodpeckers fell in 1896 in Baton Rouge, Louisiana; thousands of frogs were reported to have fallen in Acapulco, Mexico, in 1968, and spiders in Hungary in 1922.

# Most Common Place to See Waterspouts

As we mentioned previously, waterspouts like warm open water. The warmer, the better. It's that fast rise of warm air that really gets them spinning.

A great place to see waterspouts is in Florida, and even better, the Florida Keys, which are believed to be home to the most waterspouts worldwide.

**Inside the Storm**

Warm water is the key ingredient for waterspout formation.

Estimates are that some 500 waterspouts a year dance across the warm waters all the way to the Dry Tortugas.

You see, the water surrounding the keys is fairly shallow, and there is a lot of it. When water is shallow and in a warm climate, two things will happen: first, it will warm up very quickly; and second, it will be quite humid. Those water droplets from the humidity will be rising into the air faster than on Lake Michigan. And as we've learned, all that rising air feeds waterspouts.

**Eye of the Storm**

Just as Tampa, Florida, is the lightning capital of the world, the Florida Keys are the waterspout capital of the world.

Another factor is those east-west trade winds, which because of their linearity help waterspouts get moving.

Waterspouts in the Florida Keys can happen pretty much anytime, even in the morning, since the water is so warm all day long. They most frequently appear, however, in the late afternoon.

Other areas of Florida can see waterspouts, including the Tampa Bay area and Gulf Coast. They've been seen off the California coast, and even in some lakes.

Around the globe, folks may witness waterspouts in Asia and Australia, too. Think warm water and good linear clouds from trade winds.

# Keeping Clear

Don't be a gawker when it comes to waterspouts. Any storm that can flip a boat, damage a building, or suck up fish is something you only want to see in your rearview mirror.

With an estimated 20,000 boaters in southern Florida alone and millions of people visiting Florida beaches each year, waterspout safety and awareness is paramount in the Sunshine State—or anywhere.

**Storm Stats**

There are an estimated 20,000 registered boaters in southern Florida, where most waterspouts are generated.

There are no estimates of how many people die or are injured by waterspouts in a period, and hopefully not too many. Most people can get out of the way of a storm moving 10 to 20 miles per hour. It's the lunkheads who see how close they can get to it or those who greatly underestimate the funnel who are at risk.

*A waterspout hovers near Key West's Duval Street in Florida.*

*(Courtesy of the National Weather Service.)*

Weather radios are a great way to stay informed of weather hazards like waterspouts. Every boat should have one. On the beach, listen to the lifeguards and check the weather postings. If you don't have access to either, remember this: waterspouts tend to form at the beginning of a storm, such as a rain storm. You may see dark clouds with a flat bottom. That's a good indicator of a waterspout.

**Inside the Storm**

Always carry a weather band radio with you when out boating so you know what weather is heading your way.

A sailboat, for instance, may only be able to travel at speeds of 5 to 10 miles per hour, and if a waterspout forms nearby, it can mean havoc. Try to escape the waterspout by moving in a right angle away from it. If there's just no getting out of the way and you're in a little boat, it's better to dive overboard, mostly to get clear of the debris that may be spinning through it. Again, try to swim away from it. And the jury's still out whether to dive below it to escape—no one knows for certain what that water below it is doing.

## The Least You Need to Know

- Waterspouts form over warm water.
- The Florida Keys are the number-one waterspout-producing region in the world.
- Several freak occurrences—like fish falling from the sky—have been reported during waterspouts.
- The best way to keep safe during a waterspout is to steer clear of it.

Chapter 8

# Airborne Deserts

## In This Chapter

- Sand aloft
- Unnatural forces
- Black Sunday
- Worldwide sandstorm

We take for granted the dirt beneath our feet. On so many parts of our globe, what's underfoot can quickly become overhead. Sandstorms and dust storms cause great monetary damage, injury, and death every year, and it's not just in the remote regions of the Sahara and Gobi deserts. And believe it or not, improper farming and development practices halfway around the world could affect your sandstorm-free life tremendously now and more so in the future.

In this chapter, sand and dust picked up from great winds at high speeds is the focus, as well as why caution is needed to tame these sandblasting beasts.

# Terra Not So Firma

If you've never been in a sandstorm, imagine a blizzard whipping at you at 50 miles per hour, but replace that snow with sand and you get the idea.

Sandstorms are born when strong winds from a storm fetch the sand grains on the ground and push them for sometimes great distances. For the most part, the wind-swept sand is confined to the lowest 10 feet but sometimes can rise as high as 50 feet.

## Dust to Dust

Dust storms operate on the same theory, but instead of sand, it's whatever ground is available to be pulled loose and whipped aloft. Convectional currents formed by the intense heating of the ground is typically what causes the wind.

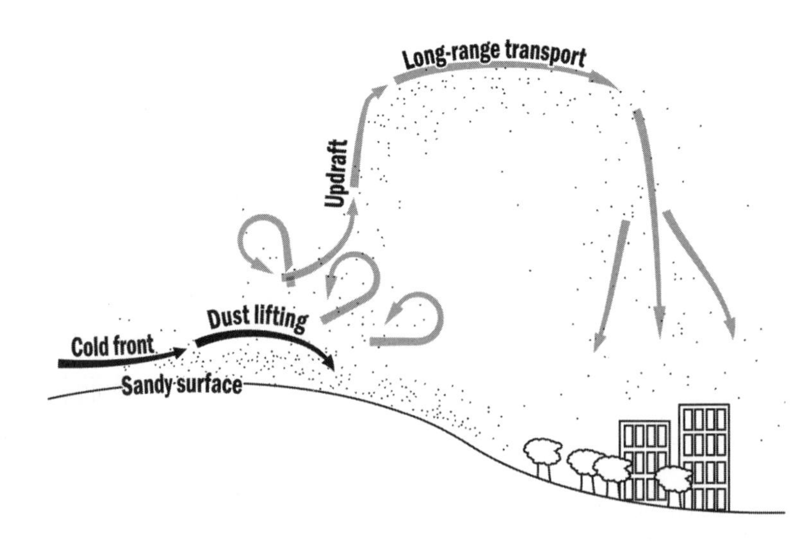


*How a sandstorm forms.*

*(Courtesy of the National Weather Service.)*

**Inside the Storm**

A haboob is a type of violent sandstorm occurring in Sudan and Khartoum that can send a solid wall of dust pluming to 5,000 feet.

Wind during a sandstorm can topple much in its path, tear up roads, snap power poles, and move dunes. Visibility in these storms often is reduced to zero, and the force of the particles can sting, choke, and even block the sun.

Such was the case in California back in 1991. A dust storm swept across a highway, causing the worst car accident in U.S. history. Many people were injured, but 17 people died in the pileup. And near Pendleton, Oregon, in 1999, six people were killed and a dozen others injured when a dust storm stacked up traffic on Interstate 84.

When the storms blow across deserts, such as during the *simooms* of North Africa, maybe only dunes are moved around. But when they hit agricultural areas, it may take years—if ever—before the soil can be restored to grow crops again.

**def•i•ni•tion**

A **simoom,** or **simoon,** is a North African and Arabian sandstorm that is so powerful it sends atmospheric dust over Europe.

Dust and sandstorms are most common in warmer months, but they've been known to whip through even in the cold of winter. In fact, a winter mix of sand and snow is called a "snuster." In 1938, a snowy blizzard mixed with dirt cut across Kansas and Texas, causing tremendous damage.

## Gone with the Wind

China has seen some wicked sandstorms over time that seem to be increasing (more on that later in this chapter). On March 10, 2004, a severe sandstorm swept through Beijing, squelching visibility to only 1,600 feet and choking and blinding residents caught in the storm.

The storm formed when a strong cold front in western Siberia pushed in, causing strong winds and blowing sand and dust over more than a million square miles.

The largest Chinese sandstorm in a decade happened in 2002, affecting more than a million residents over eight provinces in the northern part of the country. A severe drought and a cyclone in Mongolia were to blame for the storm.

**Storm Stats**

Sandstorms aren't native only to Planet Earth. In fact, sandstorms on Mars are pretty common.

Each spring, sand and dust storms pummel northern China. But more and more, the storms have shipped dust across the Korean peninsula, into Japan, and even as far as the West Coast of the United States.

But the Middle East also is home to extreme sandstorms, as much of the land in many nations is composed of sand, and the terrain is flat and hot. In fact, the storms are fairly common in the region, and for many, they're a way of life.

*Satellite image of a huge Mongolian sandstorm on April 6, 2001.*

*(Courtesy of the National Oceanic Atmospheric Administration [NOAA].)*

In January 2006, a sandstorm whipped up in the northern United Arabic Emirates, making visibility practically impossible and shutting down international airports. Winds gusted up to 30 knots while the storm moved through the region, blamed on a low-pressure system and warming conditions just days before.

## The Human Touch

Mother Nature sure can kick up some dust, but the dust has to be available for her to kick it up in the first place. Sure, many arid climates—deserts, for instance—have loads of sand for any front that wants to wreak a little havoc downwind. But the human impact on once-perfectly good grassland or forest is becoming a greater accessory, if not an impetus, to the proliferation of sand and dust storms.

**Inside the Storm**

Sandstorms form when the intense heat from the sand begins to rise, creating a difference in air pressure, and cooler winds rush in, forming a disturbance that whips up sand.

Ancient Chinese history shows that sandstorms have been documented back to reign of Emperor Zhenzong, from 960 to 1127 A.D. Songshi ("History of Song") documented that a yellow dust shrouded the sky and destroyed farm fields. Ming Dynasty (1368–1644 A.D.) writings show the devastation that a wind carrying sand brought.

With an estimated 1.3 billion people living in China, overgrazing, farming, and timbering have led to the destruction of millions of acres of land, now *desertified.*

Recent history, however, shows greater storms and greater severity, commonly covering half of China west to east. Climate does play a role in the creation of sandstorms, but data show that after the thirteenth century, the storms have been on the rise.

**def•i•ni•tion**

**Desertification** is the transformation of arable or habitable land to desert, as by a change in climate or destructive use, according to the *American Heritage Dictionary.*

There were five sandstorms in China in the 1950s, eight in the 1960s, 13 in the 1970s, and 20 in the 1990s.

Sandstorms covering more than a million square miles are happening more regularly. A storm in 1993 that was triggered by a Siberian cold front carried winds that killed 85 people, injured 264, toppled more than 4,000 houses, and destroyed massive amounts of livestock and agriculture.

The improper use of land has led to degraded fields: 34.5 percent of degradation has been caused by overgrazing, 29.5 percent by the destruction of forests, 28.1 percent by the improper use of agricultural land, and 7.95 percent by the improper use of water resources, according to a United Nations environmental report.

While China is, indeed, among the nations in most dire need of conservation efforts to ensure habitable terrain and lessen the effects of these devastating storms, other parts of the world—from Third World countries to the United States—have seen and are seeing the effects of our overambitious forefathers.

## Black Sunday and Beyond

The world was coming to an end on April 14, 1935. At least that's what folks who survived "Black Sunday" thought.

The 1930s were shaping up to be a prosperous time for farmers out in the panhandles of Oklahoma and Texas. Wheat was king, and after strong crops in the late 1920s, it looked like the economic doldrums of the Northeast's Great Depression wouldn't touch the breadbasket states.

**Storm Stats**

The longest dust storm in American history was in 1935 in Amarillo, Texas, during a record drought period that included Black Sunday. It lasted three and a half days.

But overplanting and harvesting to make a buck turned out to be the farmers' downfall. Millions of Great Plains acres were plowed, and when the huge surplus stacked up, the bottom fell out on prices and many farmers folded shop, not being able to compete. When they took down their shingles, they left the fields unplanted—barren.

Then the drought came.

## Dirty Days

Three record droughts plagued the region from 1934 through 1936. Later, record rain, floods, blizzards, and tornadoes added insult to injury, and when the windstorms came, they took with them hundreds of thousands of acres. In 1934 alone, a dust storm in Texas, Oklahoma, and Kansas blew dust clear across the nation, so that folks in New York City and Washington, D.C., could see the dusty effects. That year, extreme heat killed hundreds of people in Texas, Oklahoma, Kansas, and Colorado.

Folks thought that by April 1935 the dirty days were done. But on April 14 of that year, a giant sandstorm came rolling across the region with wind speeds recorded at 60 miles per hour. The day went down in infamy, known as "Black Sunday."

Year after year, the dust storms raged on and off. Something had to be done.

**Storm Stats**

During the American Dust Bowl years in Texas, Oklahoma, and Kansas, there were 14 dust storms in 1932; 38 in 1933; 22 in 1934; 40 in 1935; 68 in 1936; 72 in 1937; 61 in 1938; 30 in 1939; and 17 in both 1940 and 1941.

## New Practices

To say that soil conservation programs were badly needed is an understatement. New farming techniques that would protect the land would be needed if farmers were going to maintain some sort of livelihood.

The federal Soil Conservation Service estimated that 100 million acres of farmland were victim to the Dust Bowl by 1935. But some commonsense farming procedures reduced that number within five years to 22 million acres.

To remedy the losses, the federal government doled out more than $500 million to the farmers and began soil conservation programs to aid them. Conservation districts had formed by 1936 and laid down the law for maintaining strict federal farming guidelines.

The events of the Dust Bowl were so significant that John Steinbecks's classic novel, *The Grapes of Wrath*, was based on them.

# Global Sandstorm

Picture this: You're on a schooner, tropical drink in your hand, 260 miles off the coast of Florida under a gentle breeze when, on the eastern horizon, a yellowish haze starts to form. Billions of yellow crystalline particles dancing—wait, approaching—your yacht. It's not a thing of beauty, not a sailor's tale. It's sand, and it's going to get really ugly.

It happens.

## Traveling Dust

The sand plying its way across the Sahara Desert during a sandstorm can end up in your coffee cup on Key Biscayne. A Chinese dust storm can make for an odd-looking sunset in Laguna Beach, California.

How is that possible? Air currents, of course.

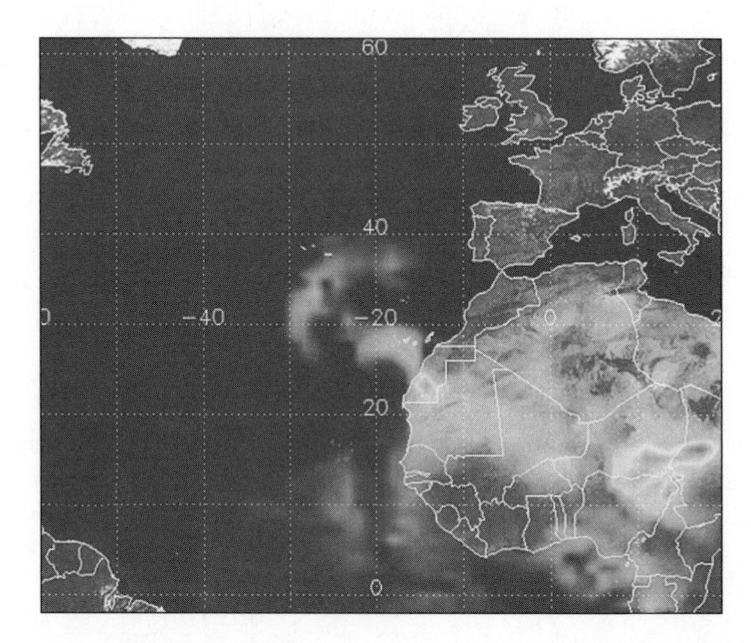


*A massive Saharan sandstorm heads across the Atlantic in this undated satellite image.*

*(Courtesy of the National Oceanic Atmospheric Administration [NOAA].)*

What's worse? The sand can have some nasty effects: breathing problems, especially for those with respiratory ailments; chemistry changes in the soil; and reflection of the sun's rays back into space.

In July, 2000, about 8 million tons—tons!—of sand from the Sahara blew all the way to Puerto Rico.

**Storm Stats**

A Mongolian dust storm in 2001 combined with Chinese air pollution and traveled across the Pacific. Particles could be seen as far as Colorado.

**Inside the Storm**

The strong southerly wind that moves from the Mediterranean Sea to the Sahara, often picking up sand, is called a scirocco.

Cool, moist Mediterranean Sea air finds its way into the Sahara every year, forcing the hot air aloft. As that hot air quickly rises, it takes a lot of dust and sand with it. It begins moving around in the upper air currents and before you know it, it's on a beeline for points west.

In July 2005, a sandstorm was heading toward Florida during the height of a spell of even hotter weather than normal. Although the weather event wasn't major, it was a reminder of how connected the planet really is.

But traveling sandstorms don't just blow off the Sahara and affect the United States. Mongolian dust storms have been known to blanket eastern Russian cities with as much as 3 inches of dust. Interestingly, the Saharan debris was carried across the Atlantic and showed up on the Eastern Seaboard about a week later.

## Reef Madness

Adverse health, climate, and soil changes are among the byproducts of a sandstorm, as we've learned, but scientists are also focusing their efforts on a place one wouldn't think sand could affect: coral reefs.

Coral reefs have recently been shown to become imperiled by sandstorms. The sands from Saharan storms that make their way to the Caribbean can carry microscopic organisms and coral reef–killing diseases.

Minerals, too, are carried in sand. Those minerals, such as aluminum, calcium, magnesium, and potassium, can enhance phytoplankton and change the delicate ocean life balance. Other species could proliferate as a result and alter the coral reefs.

## The Least You Need to Know

- China is home to some of the largest sandstorms on the planet.
- Sandstorms can kill, but are costly to national economies.
- Poor farming and foresting practices can lead to an increase of sand and dust storms.
- Extreme sandstorms can cross entire oceans.

# Part 3

# A Reason for the Freezin'

We live in some of the coldest and snowiest places in the farthest corners of the world. Year after year, the cycles of cold and snow come calling and weather records are broken only to be broken again. From extreme sub-zero temperatures to snowstorms that bury power lines and entire cities, this chapter looks inside severe blizzards, record cold and ice, and long, dark winters.

# Chapter 9

# Buried to the Eaves

## In This Chapter

- Let it snow …
- Recipe for lake-effect
- After the storm
- Extreme snowfall oddities

There's an old Yankee saying that describes nearly accurately the weather in New England: "Nine months of snow followed by three months of bad tobogganing."

It's not so different in a few other parts of the United States, or around the world for that matter. This chapter will take a look at some extreme snow accumulations and areas prone—and not so prone—to severe snow.

## How Extreme Snowstorms Form

They say no two snowflakes are alike—they are as different as human fingerprints. But when those snowflakes gather—really gather—most snowstorms are pretty similar and predictable.

But with extreme snowstorms, all bets are off.

**def•i•ni•tion**

**Snow** is precipitation composed of ice crystals in complex branched hexagonal form and often agglomerated into snowflakes.

How much can it actually *snow?* Well, folks up in the Alaskan town of Thompson Pass watched the snow fall, and fall, and fall. And fall. Over seven days in February of 1953, 187 inches of snow accumulated.

That, my friends, is an extreme snowfall.

But maybe it's not a fair comparison, either. It happened, after all, in one of the snowiest places on earth—Alaska.

But what about, say, the continental United States? Surely there must be very snowy places.

Boy, are there!

Rochester, New York, has been dubbed the snowiest large city in the United States. On average, the city receives 94 inches of snow a year. The costs to remove the snow each year runs upward of $3.7 million!

A bit west of Rochester is Buffalo, nestled among lakes Ontario and Erie. Although snowfall averages are just below Rochester's, the city saw a 39-inch snowfall in 24 hours on December 10, 1995. The cost of snow removal? A cool $5 million.

Although a good chunk of these Great Lakes cities owe their fame to snow that forms off the large bodies of water, so-called lake-effect snow (more on that in a moment), their climates and geography play a leading role in snow accumulation.

*Heavy snow blankets Taunton, Massachusetts, on January 23, 2005.*

*(Courtesy of the National Oceanic and Atmospheric Administration [NOAA].)*

## A Flaky Subject

Just about everyone has seen snow. Really. Even in Florida—southern Florida—snow has made an appearance. An inch of snow has even fallen in Phoenix, Arizona. It holds the record for the most snow ever in the desert city. It happened January 20, 1933. Folks who missed it got a second chance four years later—on the exact same date! Lightning may not strike twice, but snow sure does, and then some.

So why does it snow? There are plenty of regions around the globe that see extreme low temperatures over the cold winter months but don't get a heck of a lot of snow.

**Inside the Storm**

Snow is white because its crystals reflect most of the sunlight. Any sunlight absorbed by snow is absorbed uniformly over the wavelengths of visible light, making it look white.

If an area is particularly dry, it might not see much snow, except for freak blizzards or snowstorms. More humid areas, on the other hand, will see extreme accumulations more often.

## Two Sides to Every Snowflake

Snow is unlike rain, or other forms of precipitation. As the sun warms and evaporates water on lakes and oceans, that water changes from liquid to vapor, rising to the sky where it cools down, binds with other moisture particles, and rains. Snow is not just frozen droplets of rain (that would be sleet). Instead, the water vapor freezes into crystals, then the crystals gather and become heavy enough to fall to the ground.

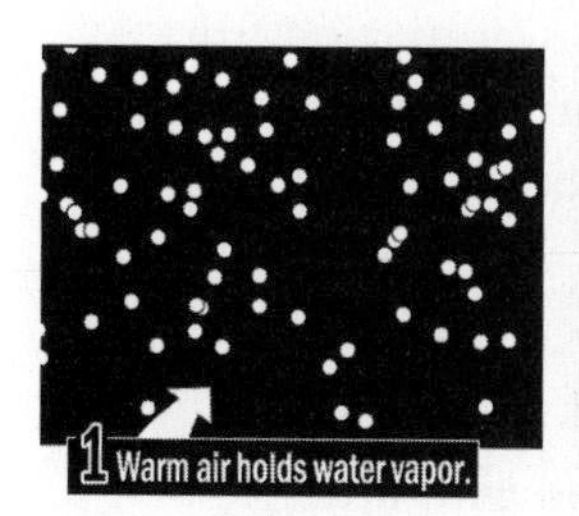


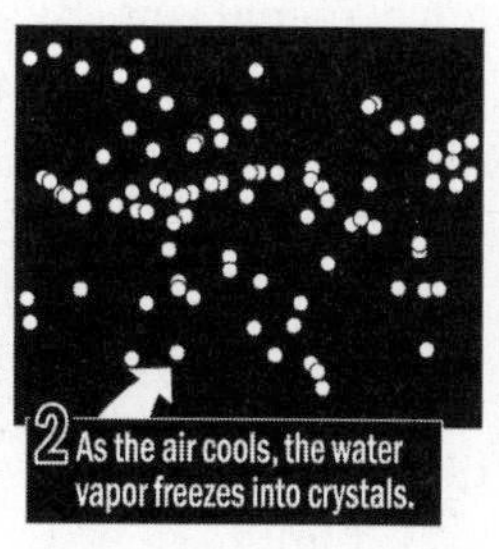

*How snow forms.*

*(Courtesy of the National Snow and Ice Data Center.)*

Snow can fall as gentle flurries—light, fluffy flakes that blow through the air without very much weight or sticking ability—or can come down in the form of a blizzard, amounting to several inches an hour.

**Inside the Storm**

Skiers around 1900 categorized snow as "fluffy," "powder," and "sticky" snow. Today, more specific terms like "champagne powder," "corduroy," and "mashed potatoes" have taken over.

Snow can happen even when temperatures are above freezing (32 degrees Fahrenheit), or when it's so cold it would seem like the snow would freeze into ice.

Each year a little more than 100 snow-producing storms hit the United States, and they last two to five days on average. It's a good thing, too: as much as 75 percent of all year-round surface water supplies in the western United States come from snow.

*A snowplow clears the roadway after a 2001 snowstorm buried parts of eastern Long Island, New York.*

*(Courtesy of the National Oceanic and Atmospheric Administration [NOAA].)*

Of course, there are downsides to the slippery stuff: accidents. Each year, hundreds of people are killed in traffic accidents due to snowy conditions. But overexertion from shoveling it off driveways or the roof and exposure from being caught in it—even being buried in an avalanche—are also related causes of snow deaths.

# Digging Out

If the conditions are right, substantial snow can fall for days on end with little reprieve. Such was the case during January 2005. On the twenty-second and twenty-third, a major winter snowstorm blanketed the Northeastern United States. By the twenty-seventh, strong winds made blizzardlike conditions. Logan International Airport in Boston measured a month-to-date snowfall of 43.1 inches, making it the snowiest month on record.

The year 2005 actually was a banner year for snowstorms. While it will surely go down in weather history for Hurricane Katrina and the relief efforts for the Asian tsunami that struck just days before the new year, let's not forget about the snow. Boston was just one example of some blustery weather that happened around the planet that year.

**Eye of the Storm**

On February 11, 1999, an amazing 57 inches of snow fell on Tahtsa Lake in the Whitesail Range of the Coast Mountains of British Columbia, which marks the snowiest day in Canadian history.

January's snow in Bulgaria was the cause of six deaths. In Greece and Albania, the heavy accumulations shut down whole villages, cutting them off from the rest of the world. Snow even fell in Rome, Italy, which is a rarity. Even more rare, the island of Mallorca, east of Spain, saw some of the snow that covered eastern Europe.

Algeria witnessed its greatest snowfall in a half-century. More than 40 people were reported injured and 13 killed as it crippled the city.

Back in the United States, nearly a foot of snow fell at Chicago O'Hare Airport; Worcester, Massachusetts, saw a whopping 51.1 inches.

Even the Piedmont area of North Carolina and Virginia saw snow. Traffic in the Raleigh-Durham area was so abysmal, more than 3,000 children were stranded overnight at their schools after the National Weather Service issued a *winter storm watch* and *winter storm warning* for the entire region.

**def•i•ni•tion**

A **winter storm watch** occurs when severe winter conditions are possible. A **winter storm warning** is issued when those conditions have begun.

Snow can be a great insulator against the cold, pump millions of dollars into the ski industry, and rejuvenate a dirty city with its clean, white blanketing effect, but it also can be deadly.

## Winter Wonderland

As we've said, in order to have snow, you need cold air and moisture. Weather forecasters look to see how cold the air will be and also how much moisture is moving into an area at the same time. Plus, the more moisture you have entering an area also helps in forming higher snow amounts. For example, if a cold front is moving across Ohio and it meets up with moisture, snow can form. However, many times that same cold front that continues to move east toward New Jersey can pick up additional moisture from the Atlantic Ocean, especially if there is an area of low pressure along the coast that pumps in Atlantic moisture. That additional moisture from the Atlantic thrown into the cold air can end up dumping a lot more snow in New Jersey than it did in Ohio.

*A devastating snowstorm pummeled southwest Arkansas during Christmas 2000.*

*(Courtesy of the National Weather Service.)*

Some seasons are snowier than others. For many people, it seems like it snowed more when they were kids than it does now as adults. There might be some truth in that. You see, the overall weather pattern has changed. With milder winters, the snow amounts are lower, and they are less frequent. The northeastern United States has been milder in recent years, producing more rain events. This milder trend can be attributed to El Niño, a shift in ocean patterns that usually brings warmer winters to the northern United States and Canada. The warmer air will bring about less snowfall and more rainfall events. Weather patterns usually change every 20 to 30 years, so chances are we will return to a snowier cycle again in the future.

## My Snowy Valentine

As we've said, places such as Raleigh-Durham, North Carolina, see 3 inches of snow and it's Armageddon. Their northern neighbors snicker when they hear of a finger's worth of snow crippling entire interstates and cities. But many of these areas aren't equipped to handle snow. There's not a $3 million snow-removal budget in place, or an available pile of rock salt. And let's face it, many folks in these parts have never driven in a snowstorm.

Between February 12 and 14, 2004, the four-state region of Texas, Louisiana, Oklahoma, and Arkansas saw major accumulations in what has been called the Valentine's Day 2004 Snowstorm.

A strong upper-level disturbance approached the region in the days before February 12, bringing with it freezing temperatures and a dose of snow.

*A north Texas town during the Valentine's Day 2004 Snowstorm.*

*(Courtesy of the National Weather Service.)*

The snow fell at an average of 1 to 2 inches an hour in much of the region, making roads and bridges impassable. As much as 8 inches fell in Taylor, Arkansas, and about 7 inches in Lewisville. Arkansas saw the greatest accumulations, but 6 inches fell in the north Texas towns of Clarksville and Maud. Broken Bow in Southeast Oklahoma saw 3 inches, a half-inch more than was recorded in Shongaloo, Louisiana.

Snow accumulations of 4 to 8 inches were common throughout the entire region, and most of it fell within three to four hours.

The good news was that most, if not all, of the snow melted the day after the storm, and temperatures reached back into the 50s under sunny skies. The bad news was that a lot of Valentine's bouquets didn't get delivered ....

**Storm Stats**

Valentine's Day, 2004, in Amman, Jordan, also was snowy. Parts of the city witnessed 2 feet of snow and residents built snowmen in front of the parliament building.

# Lake-Effect Snow

Let it snow, let it snow, let it snow. They're not just the lyrics of a holiday song; it should be the anthem for many of the Great Lakes cities and other areas subject to lake-effect snow.

While many of these cities are prone to cold and snowy winters even without this weather phenomenon simply due to their northern climates, insult is added to injury whenever those cold Canadian winds blow down across the Great Lakes—Michigan, Superior, Huron, Erie, and Ontario—dumping more snow, even blizzardlike amounts.

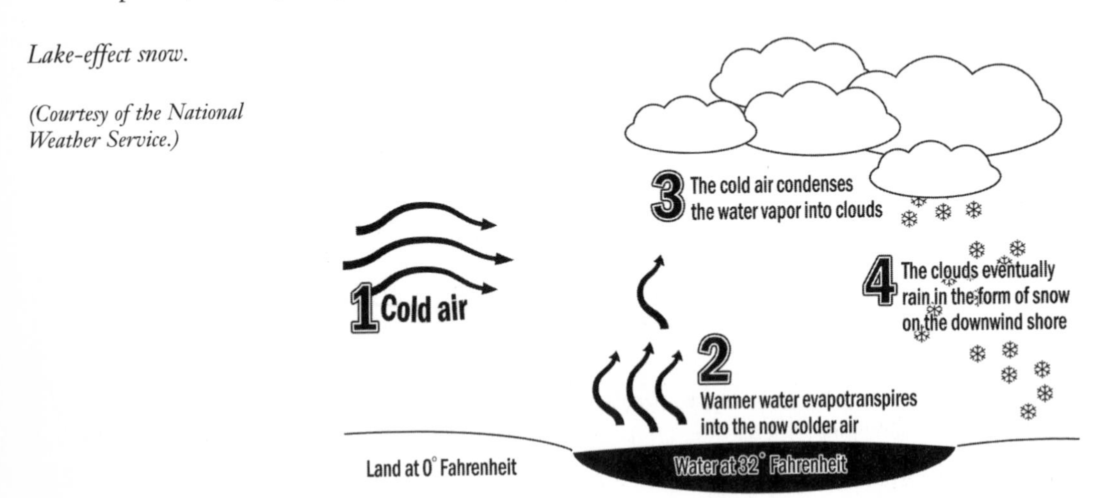


*Lake-effect snow.*

*(Courtesy of the National Weather Service.)*

Here's how it works: when cold Arctic air sweeps across the Great Lakes in the winter, it mixes with the relatively warm water off the lakes as it makes its pass from the north or northwest. The cold air is quickly warmed from below, water evaporates, and the small particles of moisture freeze as they rise into the colder upper atmosphere. Those particles band together, create heavier snowflakes, and fall back to earth accompanied by the strong winds from the pressure difference between the warm and cold air.

**Inside the Storm**

Lake-effect snow is an isolated weather event. While one city is being hit with this type of snow, another city several miles away could be enjoying a sunny day.

Even central New York cities such as Rome, which isn't exactly on the shores of Lake Ontario, but close enough, get pummeled with lake-effect snow.

Lake-effect snow systems often deliver a one-two wallop. The snowfalls don't just occur from the cold air passing over the warm lake, but when that snow is pushed with the force of the wind differential between the low and high pressures. For two to three days following that cold passage, even colder and more blustery weather occurs.

And because the weather event is prone to common winter patterns—northwesterly winds blowing across the lakes—once one system passes, another can begin, even within a half-day of the last squall.

## Unwelcome Encore

Tom Dudzick, a playwright born in Buffalo, New York, wrote a play that was canceled because of a massive lake-effect snowstorm that slammed into the city from December 24 to 28, 2001. More than 82 inches of snow fell during the four-day storm.

The title of Dudzick's play, ironically, was *Lake Effect.*

Dudzick told *The New York Times,* "The phrase 'lake effect' sure seemed a lot funnier last summer when I came up with it for the title."

No kidding.

All but 1 inch fell for the month during that four-day span. The record for snowfall was in 1985, also in December, when just over 68 inches fell.

**Inside the Storm**

Higher elevations east of Lake Ontario receive more than 200 inches of snow annually, making that area the snowiest populated region east of the Rocky Mountains.

*NOAA's National Weather Service Forecast Office in Buffalo, New York, on December 29, 2001. The governor had to call for a state of emergency and the National Guard came in to help uncover the city. Snowplows from nearby cities even pitched in to help. The month of December in Buffalo that year saw 83 inches of snow.*

*(Courtesy of the National Oceanic and Atmospheric Administration [NOAA].)*

### As Great as the Great Lakes

Although the Great Lakes shorelines bring great snowfalls, lake-effect snow isn't confined simply to this large region. Any large water body with enough *fetch* can yield lake-effect snow. The Great Salt Lake in Utah produces lake-effect snow, as do the eastern shores of the Hudson Bay.

> **def•i•ni•tion**
>
> The wind must sweep over at least 50 miles of lake surface before lake-effect snow can happen. That distance is called the **fetch**.

The Great Salt Lake in Utah adds salty snow each year to Salt Lake City.

Scandinavia sees regular bouts of lake-effect snow, as do Korea in the ocean-effect snow from the Yellow Sea and Hokkaido from the Sea of Japan.

## Freak Snowstorms

Snowball fights in Texas? Snow angels dotting schoolyards in San Diego? Snowmen lining the streets of South Florida?

Well, let's not get too crazy. But the fact of the matter is that each of these areas has seen the white stuff fall from the sky during winter months.

It's odd, indeed.

### Snow in South Florida

Although the Florida Keys have never seen snow, a half-hour's drive north of the mainland bridge has.

Homestead, Florida, reported flurries on January 19, 1977, during one of the harshest cold waves of that decade. Although no official measurement exists, the event has made it into the weather books.

In Miami, on the same day, residents bundled up against the bitter cold and watched as snow—snow!—fell on Miami Beach. It was the first time snow had ever fallen in Miami. Headlines on the 20th in the *Miami Herald:* "Snow Falls on S. Florida."

> **Eye of the Storm**
>
> High school softball games were canceled on February 14, 2004, in Dallas, Texas, when 3 inches of snow fell on the city.

Not the punchiest headline, but it was written in the WAR font.

It was a cold month for the nation, with a flow of Arctic air buzzing straight down toward Florida. A mid-January nor'easter spun northward from Hatteras and pushed cold Hudson Bay air as far down as the Caribbean. The events set up the perfect conditions for snow, and on January 18, Florida Panhandle residents witnessed as much as 2 inches of snow. Even Tampa, Orlando, and Vero Beach reported flurries.

The cold pushed on toward West Palm Beach, and on January 19, a forecast of snow was issued for the first time in Miami.

And the prediction was right on.

## The Frozen Tundra of San Diego?

Hell didn't freeze over, but residents of San Diego, California, may have thought so when they witnessed snow falling in January of 1882. Dubbed The Great Storm of January 1882, snowflakes fell on the Southern California city, but were melting as soon as they hit the ground.

However, 3 inches were reported 18 miles east of downtown, in El Cajon Valley. And the town of Campo witnessed 3 feet of snow on that day. Of course, Campo is a mountain town, roughly 30 miles from San Diego, but it already had been snowing for three days ….

But 1882 wasn't the last time that snow fell in San Diego's city limits. Flurries were reported at Lindbergh Field in January 1937; snow fell again in January 1949 (3 feet at the 6,000-foot level of Mount Laguna, 45 miles east of San Diego); in 1967, a SoCal storm brought snow to parts of San Diego, such as Del Mar, Encinitas, Vista, and La Jolla.

Not too shabby.

## The Least You Need to Know

- Snow can fall just about anywhere that temperatures and moisture are right—even in Phoenix.
- Lake-effect snow is an isolated snow-producer most common to Great Lakes shore cities.
- Snow can fall in odd places—such as Miami and San Diego!

Chapter 10

# Blizzards

## In This Chapter

- Ingredients for severe snow
- When wind and snow meet
- A winter of blizzards
- 1888's banner blizzard

Funny name, not so funny storm. No one really agrees on where the name "blizzard" comes from, and around the globe, there are many different versions or names altogether.

Early American settlers used *blizz* to describe the wind-driven snow, and the Germans used *blitartig* when referring to sudden snowstorms.

Russians have their *buran,* Siberia has the *purga,* and the French in the southern part of the country call it *boulboe.*

Call them what you will, blizzards might be great for ski areas, but to humans, they can be deadly. In this chapter, we'll show how blizzards form, look at areas that are most prone to these severe storms, and review some of the worst blizzards of all time.

# How a Blizzard Forms

The jet stream has much to do with *blizzard* formations. When the jet stream's cold Canadian air dips south, it meets warmer air and fronts clash to form a storm.

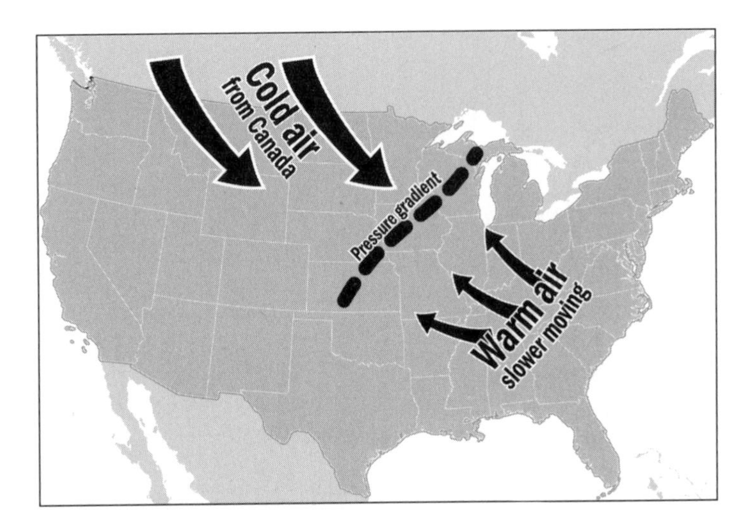


*Pressure gradient.*

*(Courtesy of the National Weather Service.)*

Blizzards usually form on the northwest side of the storm system, when the difference between high- and low-pressure systems creates a tight pressure gradient, literally squeezing strong winds out of the system. Add snow to the mix, and a blizzard is born.

**def•i•ni•tion**

A **blizzard** is a snowstorm with 35-mile-per-hour-plus winds with visibility less than a quarter-mile for three hours. It's a severe blizzard at 10 degrees Fahrenheit with winds exceeding 45 miles per hour and zero visibility.

There are different blizzard regions across the United States; where and how they typically form varies.

Generally, when we think of the classic blizzard, we're talking about the *nor'easter.*

**def•i•ni•tion**

A **nor'easter** is a powerful low-pressure system that forms when the warm seawater meets the cooler late-season air over land, affecting the Mid-Atlantic to New England with snow, wind, and rain.

*Three men try thawing switches to keep the streetcars running during the "Knickerbocker" storm of January 27–28, 1922, in Washington, D.C.*

*(Courtesy of the National Weather Service.)*

Mountains, valleys, and other low elevations tend to trap cold air—cold, low-pressure air is "heavier" than warm, high-pressure air. When the warm air moves over the cold, remember, pressure flows from high to low, the rain falls from the warm through the cold and freezes. Although the southern Atlantic and Gulf Coast states are usually immune to these snowy systems, periodically, cold Canadian air moves far south with the jet stream, combining with the warm Gulf air to pummel the region with a blizzard.

**Inside the Storm**

When warm, wet air passes above cold, dry air, precipitation results. When the rain droplets pass through the cold, they can turn to snow.

March 12, 1993, is a good example. On that day, a low-pressure system was building over the Gulf of Mexico and starting its trek northeast. Residents all over Alabama hunkered down when winds as strong as 55 miles per hour combined with snow hammered the state through the night. Every part of the state had been blanketed by snow, much of it more than a foot. The region, not used to—or prepared to deal with—blizzards, was paralyzed. In the storm's wake, 14 people were dead due to exposure and an estimated $50 million in damages occurred.

Midwestern and Plains states see blizzards come from a different path. These storms form over the American or Canadian Rocky Mountains and southeastern Colorado and move northeast, again with cold Canadian air mixing with warm gulf moisture. Wind from the Canadian Rockies can plunge temperatures as low as 70 degrees below zero Fahrenheit!

*A man stands beside power lines buried by a blizzard in Jamestown, North Dakota, on March 9, 1966.*

*(Courtesy of Dr. Herbert Kroehl, NGDC, NWS.)*

And the West Coast isn't without its blizzards, either. North Pacific storms pummel the coast with moisture; when one moves inland and meets the cold, trapped air of the mountains, a blizzard is born. When those fronts clash in a mountainous region, the wind whipping through the canyons can create their own intense pressure, and can reach an excess of 100 miles per hour.

In Alaska, blizzards are the norm. But what degree of blizzard? Try wind chill temperatures at 90 degrees below zero Fahrenheit! Temperatures of 40 to 60 below can last a week, and wind-driven waves can drive massive ice into coastal towns, tearing down buildings and claiming lives.

## Whiteout

Legend has it that South Dakota holds the title for the most blizzards in the United States. Although data are sketchy, even today, the facts add up: from 1978 to 1999, South Dakota has seen 265 blizzards. The mean for the nation is about 12 per year. In 1981 alone, South Dakota dug out from 28.

The winter weather was so severe, South Dakota was dubbed "The Blizzard State." Realizing it wasn't the best marketing concept, the state then dropped the nickname.

But areas prone to blizzards are used to faring better in them than those places that don't often see or aren't well-equipped to deal with the storms. We're not talking about a powerful storm in otherwise temperate Alabama, but the Blizzard of 1996 sure is a good example.

December 1995 was shaping up to be a typical eastern winter, with several squalls, snowstorms, and blizzards. Then 1996 roared in like a lion, dropping as much as 4 feet of snow over the Appalachians, Mid-Atlantic, and Northeast. But this system didn't contain itself to one region. Many states across the United States had been affected. When the snow had settled, 187 people were dead and damages, also due to the residual snow melt and rain, totaled approximately $3 billion.

Snowfall amounts totaled 23 inches in Elkins, West Virginia; 27 inches in Newark, New Jersey; 30 inches in Philadelphia; and 14 inches in Cincinnati. All were state records.

But as the cold weather system spread throughout the South and Midwest, the death tolls began mounting. Kentucky saw six dead, while Alabama lost four. Thirteen Virginians lost their lives in the blizzard, and in Pennsylvania, the hardest-hit state, 80 deaths were reported. Deaths from the snowstorm alone totaled 154.

**Storm Stats**

During the Blizzard of 1996, a whopping 27.6 inches of snow fell on Philadelphia in 24 hours.

A sudden warm-up followed the blizzard, and flooding was rampant. Temperatures climbed to 65 degrees in Burlington, Vermont, on January 19. The Delaware, Susquehanna, upper Ohio, Potomac, and James river basins crested to as much as 30 feet above flood stage, killing 33 people and leaving an estimated 200,000 homeless.

There are perhaps hundreds of dramatically named severe blizzards throughout the country and around the world. In each case, massive amounts of snow and driving wind catching a city off-guard seem to be the ingredients for dubious-titled storms.

One in particular was the "Storm of the Century" on November 25 through 27, 1950, in the eastern United States. Heavy snow and wind across 22 states claimed 383 lives, and damages totaled an estimated $70 million. Just a year before, on January 2 through 4, 1949, Nebraska, Wyoming, Utah, Colorado, Nevada, and, of course, South Dakota, witnessed 72-mile-per-hour winds that formed snowdrifts 30 feet tall. The sudden storm, bringing less than 3 feet of snow to most parts, destroyed tens of thousands of cattle and sheep.

**Eye of the Storm**

Midwestern storms can sweep across the plains at speeds so fast that livestock can be trapped. Cattle have been found frozen solid, still on their feet, months after a blizzard—during the spring thaw.

In 1977, a cartoon in *The Buffalo News* showed an airplane over a perfectly white background. The pilots were radioing in that they were flying over Buffalo, New York. Seven inches of snow were added to the 35 inches already accumulated during the January 28 through 29 Blizzard of 1977. Wind gusts drove upward of 70 miles per hour with snowdrifts reaching 30 feet high. Twenty-nine people were killed in the storm and seven counties in western New York State were declared national disaster areas.

New England was battered by the Blizzard of 1978 on February 6 through 8. Two to 4 feet of snow fell and 54 people were killed. An estimated $1 million in damages were incurred. In 1993, 270 people lost their lives in the Superstorm of 1993 from March 12 through 14 on the eastern seaboard. Damages were believed to have reached a staggering $6 billion when snow fell at a rate of as much as 3 inches an hour.

*A young steer tried to defrost after a March blizzard in Rapid City, South Dakota. Blizzard conditions are extremely hard on exposed livestock.*

*(Courtesy of the National Weather Service.)*

The lists go on and on through history, with stories of travelers starting out on normal days only to be trapped in a blizzard hours later. Or schoolchildren caught in a tempest on their way home from class.

But what about the neighbors to the north or around the world? Surely they must have record blizzards and their own tales of extreme winter weather events.

Boy, do they!

In the heart of Canada's prairies, namely Alberta, wicked blizzards have engulfed many towns and cities. On December 15, 1964, Alberta and the southern portion of Saskatchewan fell victim to extreme wind and cold from a blizzard. Snowfall was relatively light, but minus 22 degree Fahrenheit temperatures and 46 mile per hour winds killed thousands of cattle. Three people died when their stoves went out during the night.

**Storm Stats**

During the 1971 Storm of the Century, work crews hauled away 500,000 truckloads of snow from the Montreal streets.

On New Year's Day, 1973, the Peace River and Fort St. John areas were plagued when a wicked blizzard pounced on the region. A crew was killed when their 707 crashed upon landing near Edmonton. And during a two-day blizzard on January 29 through 31, 1989, seven people died in Edmonton when cold Yukon air bore down through Calgary, with temperatures as low as minus 13 degrees Fahrenheit and snow accumulation totaling nearly 14 inches.

From January 30 to February 8, 1947, a 10-day blizzard buried towns from Winnipeg to Calgary and closed railways until spring. Snow was so high that folks were stepping over power lines; accounts have it that a farmer in the Moose Jaw area had to cut a hole in the roof of his barn to get to his cows.

On March 4, 1971, Montreal, Quebec, suffered its own Snowstorm of the Century. All told, 17 people died when 18 inches of snow driven by 68-mile-per-hour winds fell on the city. Electricity in some parts of the city was out for 10 days.

## The Blizzards of 1888

Two significant blizzards struck the United States in 1888. While The Great White Hurricane, as it was called, pummeled the Northeast from March 14 through 18, the year got off to a bad start on January 12, when the Dakota and Montana territories, along with Minnesota, Nebraska, Kansas, and Texas, were hit with such a powerful storm system that 235 people were left dead in its terrible wake. Sadly, most of the deaths were children, caught in the

**Storm Stats**

How famous was the storm? Laura Ingalls Wilder even wrote about the great blizzard in her children's book, *The Long Winter.*

storm as they walked home from school. The weather event had been dubbed the Schoolchildren's Blizzard.

But just a few months later, the most famous blizzard of all time struck the East Coast from the Chesapeake Bay to Maine.

Days before the storm hit, heavy rain began falling along the East Coast before changing to snow when the mercury dropped on March 12. For the next 36 hours, the snow did not cease.

Winds picked up to nearly 50 miles per hour and 50 inches of snow fell in Massachusetts and Connecticut and 40 inches in New York and New Jersey.

In Boston in 1888, storm predictions weren't terribly reliable. And the Great White Hurricane wasn't forecast. It trapped hundreds of railway passengers in trains, and those who did make it home were stranded in their own homes for as much as two weeks. Many people froze, many others starved to death.

When the storm had ended, 200 ships were grounded and 100 of their seamen were killed. An estimated $25 million worth of property damages occurred. The death toll from this violent blizzard was reported at 400.

Planners and leaders in the metropolises affected by the storm made their cities safer, burying power and telegraph lines, building underground subway systems, and using weather balloons to better forecast storms.

## Surviving the Storm

Consider this: most blizzard-related deaths happen in automobiles—70 percent. About 25 percent of fatalities in a blizzard are due to people being caught in the storms. Keeping warm, dry, hydrated, and unburied is the way to beat a blizzard, and with a few proper actions, your chances of survival if caught in a blizzard increase significantly by knowing what to do.

Dehydration is one of the main killers in a blizzard. The human body needs water—and actually will stay warmer with water—to survive in cold weather, too. Luckily, there is a wealth of water during a blizzard: snow! But don't eat the snow—it will lower your body temperature. Instead, melt the snow first, then drink it. Food also produces energy, which produces heat.

Dressing in layers will also increase your chances of staying alive. Loose-fitting, lightweight clothing keeps perspiration at bay. And, when overheating during exertion, removing layers is easy. Sweat plus cold equals a good chill. It's important to avoid that. Also, every part of your body should be covered.

## Stuck in the Car

If there's no shelter nearby and you're caught in your car or truck, stay inside. When the wind and snow are driving, it's easy to become disoriented and lose your way. The key is to roll down the window a little to keep a supply of fresh air and to prevent carbon monoxide from building up. Running the motor for about 10 minutes every hour will allow you to run the heater periodically. The exhaust pipe must not be blocked.

You should make your car visible to rescuers. A flag or a cloth on your antenna will help. Also, periodically move your legs, arms, fingers, and toes to keep the blood circulating and to keep warm. Try not to build up a sweat, though.

Keep your vehicle fully winterized, with a full tank of gas, and have these items handy: blankets and sleeping bags; flashlight with extra batteries; first-aid kit; knife; high-calorie, non-perishable food; extra clothing; a large empty can and plastic cover with tissues and paper towels for sanitary purposes; a smaller can and water-proof matches to melt snow for drinking water; sand; shovel; windshield scraper and brush; tool kit; tow rope; jumper cables; and water.

## Home with No Heat

If at home when a blizzard hits, and the power and/or heat goes off, use alternate heating sources, such as a fireplace, woodstove, or space heater. Be sure the space heater is approved for indoor use and ventilate properly. Keep all combustibles away from the heat source. It also would do you good to seal off unneeded rooms by closing doors and putting rags or towels in the cracks under doors and over windows at night.

Have these items handy: a flashlight, extra batteries; cellular phone; battery-powered weather radio; high-energy prepared food; 1 gallon of water per person per day; medicine and baby items, if needed; first-aid supplies; heating fuel; and an emergency heating source.

## Building a Shelter

In the worst-case scenario, you're going to find yourself outside and not in a car or near any shelter. If that happens, you're going to need to build a shelter to survive.

First, you need to get out of the wind. A lean-to, wind-break, or snow cave will do the trick. You'll also need a fire to keep warm and attract attention. Rocks around the fire will soak up and reflect the heat.

According to the U.S. Search and Rescue Task Force, building a snow cave takes a shovel and a few hours of work, but it could save your life. Sweat and moving snow also will soak your clothes, so it's better to have a change of clothing. It's also good to split the duties with someone else to avoid fatigue and perspiration.

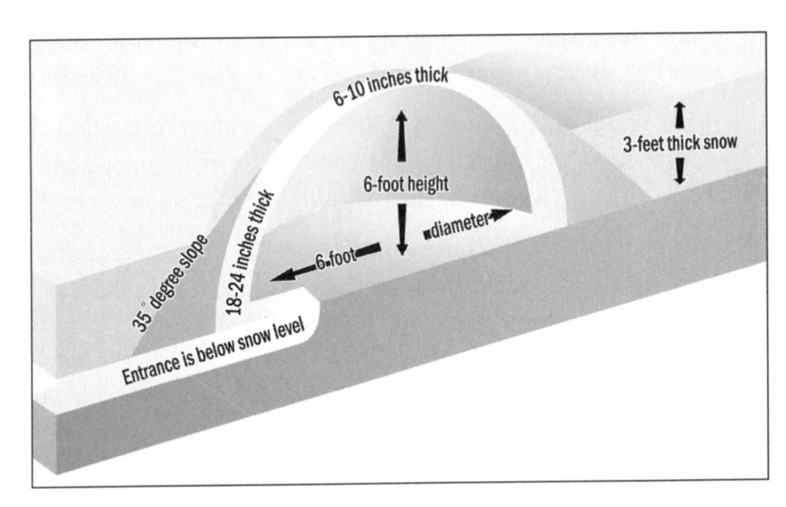


*Dimensions of a snow shelter.*

*(Courtesy of the USSRTF.)*

*Making a snow cave.*

*(Courtesy of the USSRTF.)*

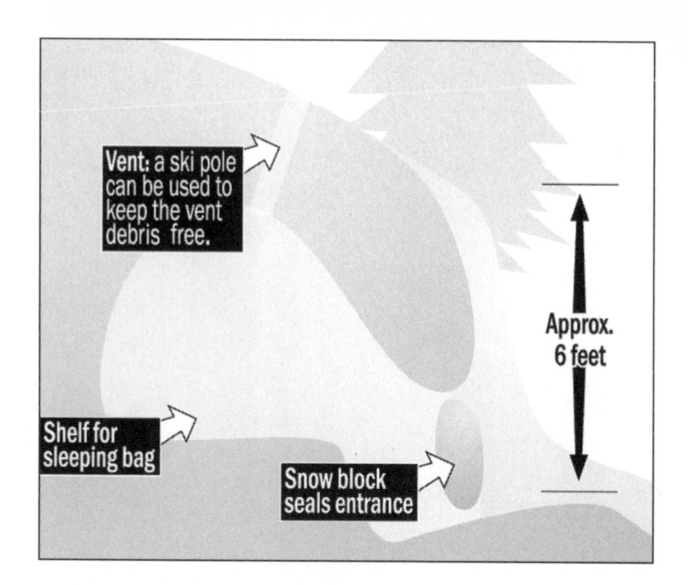


*Snow cave on a slope.*

*(Courtesy of the USSRTF.)*

Blizzards are unpredictable, blinding, and can be as violent as other extreme storms. Preparing for the worst could mean the difference between life and death.

## The Least You Need to Know

- Blizzards are common around the world, but extreme blizzards have killed thousands and racked up billions of dollars in damages.
- The Blizzard of 1888 is the most famous of all time.
- Preparation is the best way to beat a blizzard.

Chapter 11

# The Big Slip: Ice Storms

## In This Chapter

- Slippery when wet (and cold!)
- Ice jam
- Southern inhospitality
- Extreme Canadian ice storm

Driving in an ice storm is nuts, but so many of us have done it—and continue to. Heck, just walking on ice caused by a frozen rainstorm is tough enough.

But driving or walking during extreme freezing rain? Now we're talking about taking cover, because power lines fall and arc, tree branches snap like twigs under the weight of frozen rain, and cars become one-ton hockey pucks on ice-slick roads.

In this chapter, we'll clear up some myths about how rain freezes, and look at ice storms that have turned whole cities into giant ice-skating rinks.

## Why Does Rain Freeze?

Precipitation is a tricky thing. Warm, summer rain can be cleansing, nourishing; winter's frozen rain and sleet, however, can be devastating.

But while these two frozen elements may yield the same outcome—losing your feet from under you—their journey to earth begins as two distinct opposites.

## Freezing Rain vs. Sleet

*Freezing rain*, contrary to popular belief, as well as its name, is made up of supercooled droplets of moisture that freeze when they hit the cold ground. The drops of rain are in a liquid state as they make their earthbound journey.

Here's how it happens. Typically, freezing rain is found on the north side of a warm front, usually in a thin band. It actually begins as falling snow high up in the atmosphere where it meets the front's warm air and melts into rain. As it falls, it sails back through freezing temperatures (remember, warm air rises) but doesn't freeze. The process is called supercooling. When the supercooled droplet smacks your frozen sidewalk, it freezes and you have frozen rain to scrape away, or it bears down on your pine tree, the power line in front of your house, or, heaven forbid, the windshield of your car as you creep along the interstate.

**def•i•ni•tion**

**Freezing rain** is moisture that remains in a liquid state until making contact with a frozen surface. **Sleet** is moisture that falls to earth already in a frozen state.

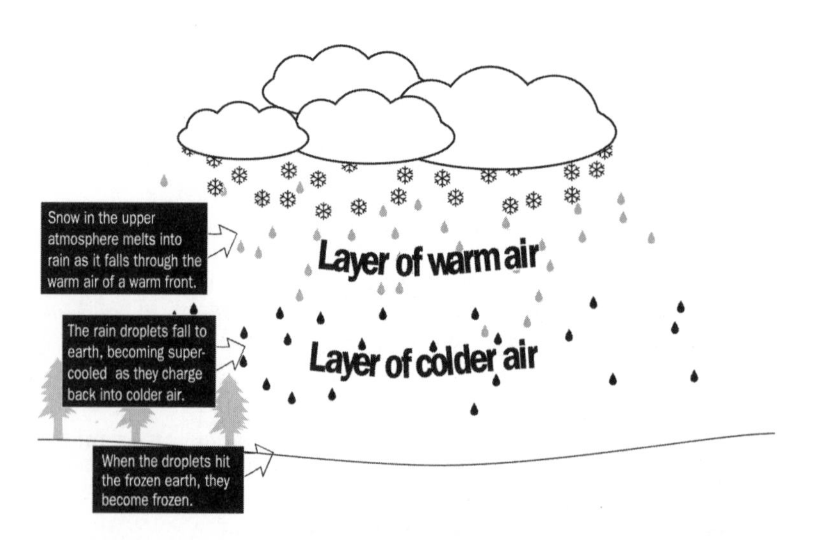


*How freezing rain forms.*

*(Courtesy of the National Weather Service.)*

*Sleet*, on the other hand, looks a lot like freezing rain when you slip on it, but there are a few differences. First off, while sleet pretty much starts its earthward journey as snow, but sometimes rain, the temperature that it falls through is cold enough to freeze it solid, making it come down as frozen (and often stinging) pellets.

It's possible for sleet to change to freezing rain, and vice-versa. In the course of a day, snow also could be possible, making for what weather experts call a wintry mix. It's possible that sleet and freezing rain could happen simultaneously as one form changes over to the next.

## A Slippery Slope

Ice storms are generally categorized as an accumulation of a quarter inch or more. It usually takes a couple of hours for that to happen.

While freezing rain is pretty common in the United States, ice storms aren't. Most freezing rain falls from New York into Virginia, and on the West Coast, Washington and northern Oregon, but freezing rain has occurred as far south as Florida, Texas, and even southern California.

And the frozen stuff can be deadly. From 1990 to 1994, ice storms were the cause of 10 deaths, 528 injuries, and $380 million in damage, according to the National Weather Service.

**Eye of the Storm**

In January 1998, much of the Northeast was plagued by an extreme ice storm that killed 16 people and caused more than $1.4 billion in damages.

But in January and February of 1951, an estimated 25 people died and 500 were hurt when temperatures in parts of Kentucky plummeted below freezing and mixed with a warm front that caused a massive ice storm. In the history books, the weather event is known as The Great Ice Storm of '51.

A high-pressure system pushed cold air into the South at the same time a cold front pushed in from the Gulf of Mexico all the way to New England, and on January 31, the precipitation began falling. In Nashville, Tennessee, temperatures were at 17 degrees Fahrenheit. Temperatures at 5,000 feet, however, were close to 50 degrees. As we've learned, when snow melts in the upper atmosphere and falls on frozen ground, ice forms.

And form it did. In Bowling Green, Kentucky, 3 inches of precipitation fell and covered the area with an icy crust. On February 1, the freezing rain again fell, with air temperatures hovering at zero degrees. On February 2, minus 20 degrees was recorded in Bowling Green and negative 13 in Nashville. Whatever was on the ground froze solid. Power lines snapped, trees and limbs fell down throughout the icy region, pipes froze solid, trains fell days behind schedule, and the airports were closed for three days.

**Eye of the Storm**

Without power during extreme cold, Nashville residents expended the city's entire fuel supplies during the three days of The Great Ice Storm of '51.

All told, losses were estimated at $100 million and it took more than 10 days for the area to return to near normal. Livestock, agriculture, and forestry suffered great losses.

In Nashville, more than 16,000 homes were without power and transportation and communication damages were estimated at $2 million.

# Ice Jam

All that ice has to go somewhere. Unlike rain, which percolates into warm soil, assuming the soil's not completely saturated, freezing rain remains in a solid state until a thaw—sometimes at the end of winter—and begins its gradual melt. Even during gradual melts, flooding is often the byproduct. In more temperate climates, however, ice from a freak ice storm melts quickly and can swell rivers over their banks in the blink of an eye.

Also causing trouble on the rivers are *ice jams.*

*Ice chunks float along the Wenatchee River in Washington in this undated photo.*

*(Courtesy of the U.S. Geological Survey Spokane Office.)*

In northern Maine, warm spring temperatures tend to send ice and snow runoff into the St. John and Allagash rivers, swelling the water bodies to 2 feet above flood stage. In April 2005, several ice jams formed in the rivers as the Allagash alone rose to nearly 20 feet, flooding homes and roads and damaging bridges in the North Maine Woods.

On the St. John River, an ice jam backed up for 12 miles, stacking up ice as high as 30 feet!

What's so critical about an ice jam? That ice eventually melts, and if it's all bunched together in one place, the flooding can be catastrophic, not to mention the extra masses pushing already high waters over their banks.

**def•i•ni•tion**

Broken river ice that becomes trapped in a channel is an **ice jam**. The result is usually flooding.

Just think of a bathtub filled to the brim. Add your two feet and ankles, and the water goes over the sides.

Northern Maine's rivers have seen significant ice jams throughout time. In December 1990, mild temperatures broke up the St. John River ice and then it refroze several times, piling the ice jam three stories high near the little village of Dickey by April 1991. By April 9, a 740-foot bridge over the St. John was overtaken by the ice flow, which twisted it off its foundation. The ice claimed another footbridge on the Little Black River and when it pushed over the banks of the town of Allagash, 30 homes were destroyed. Aroostook County was declared a federal disaster area.

In 1999, the first week of January was grueling for the entire eastern United States and Canada. Ice storm–related deaths totaled 56—29 of which were in Canada. More than 3 inches of frozen rain fell on upstate and northern New York into Maine and Canada, where 3 million customers were without power. About 500,000 New England customers were without electricity, and about 80 percent of those residents were in Maine.

## Southern Discomfort

The South—especially the Deep South—of the United States is known for its warm climate, and central and southern Florida have wonderful 80 degree days in the middle of January. But don't bet on never seeing ice in Miami Beach or Orlando, because it's happened.

Doing donuts in a pickup truck in the iced-over parking lot of the Daytona Beach Wal-Mart isn't a common occurrence, but the northern Gulf of Mexico side of the state, along with Mississippi and Alabama, have seen their share of crystal-froze water.

Low-pressure systems occasionally sweep across the Gulf, which can pick up massive amounts of moisture over the temperate waters and dump it over these regions with ferocity.

Back in the winter of 1973, Atlanta, Georgia, was hit with up to 4 inches of ice. January 7 and 8 temperatures remained at about freezing.

In Florida, Jacksonville's white Christmas in 1989 began with an ice storm that brought freezing rain to the region on December 23. And in the Carolinas, one man was killed when a tree, burdened by ice, fell through his Charlotte, North Carolina, house in 2005.

Nearly 2 million Carolinians were left in the dark when the December ice storm wreaked havoc on the region. Georgia and Virginia also were affected when the freezing rain started falling on December 15. In its wake, 24 people were killed, power lines were downed, and trees snapped like twigs.

**Storm Stats**

The flexibility, tapered shape, and lack of branching enables conifers to sustain heavy snow. During a severe ice storm, a 50-foot-tall conifer can hold 99,000 lbs of ice.

The storm was the worst the Carolinas had seen in years. Roughly 3,000 travelers watched as their flights were canceled at Charlotte-Douglas International Airport and traffic on the area's roads was treacherous.

Texas has seen its share of ice storms, too. President Bill Clinton declared Texas a federal disaster area after a series of crippling ice storms plagued the region beginning December 12, 2000.

*Texarkana and Oklahoma were plagued by an ice storm on December 25–26, 2000.*

*(Courtesy of the National Weather Service.)*

Another storm swooped in on Texas on December 25 and 26. The Christmas was white in northwestern Texas, but it was harrowing, too. The ice and snow mixture accumulated to as much as 4 inches by the twenty-sixth. An estimated 25,000 residents lost power, schools were closed, and damages totaled $200,000.

But the Lone Star State wasn't alone in the fight against the blustery weather. Oklahoma took the brunt. As much as a foot of snow began falling on Oklahoma on the twenty-sixth, but a few miles east, sleet and freezing rain were the rule—as much as 8 inches of mixture in Enid and Weatherford. It was southeastern Oklahoma's worst ice storm ever, with sleet and freezing rain adding up to 2 inches in parts and an inch of ice covered Cotton County.

Electricity lines were severed and residents were without power for as much as a week. The weather-related death toll was six, mostly from automobile accidents, but a utility worker was electrocuted trying to restore power.

**Eye of the Storm**

A whopping $76 million in damages occurred in Oklahoma alone during an ice storm that stretched across northwest Texas and above on Christmas, 2000.

# The Canadian Ice Storm of 1998

It was the most costly natural disaster in Canadian history. As much as 4 inches of ice formed a deadly crust over Ontario, Quebec, and New Brunswick for nearly a week in January 1998. The extreme weather event went into history books as estimated costs of storm damage crested at $5.4 million (Canadian) and insurance claims reached the $1 billion mark (Canadian).

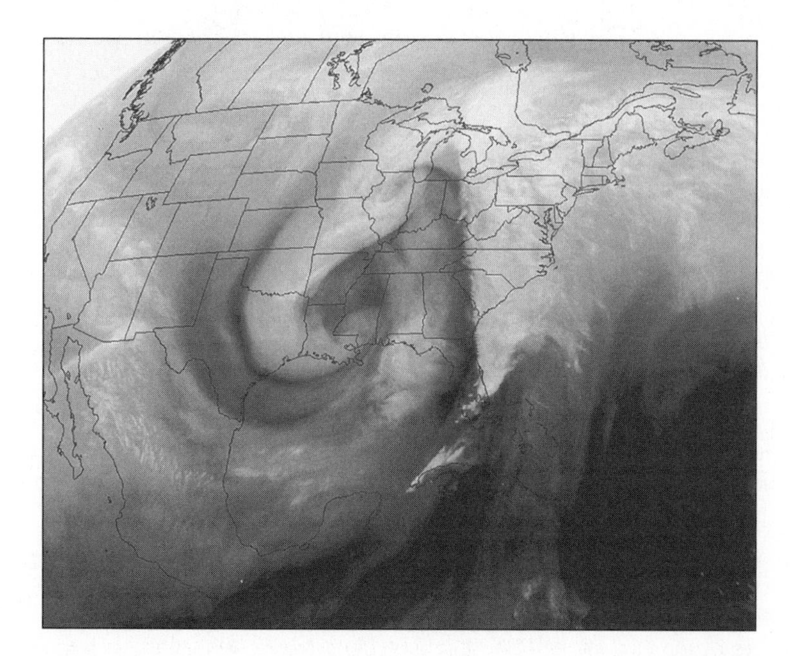

*Satellite image of the ice storm of 1998.*

*(Courtesy of the National Oceanic and Atmospheric Administration.)*

> **Storm Stats**
>
> Freezing rain during the January 1998 ice storm fell for more than 80 hours—nearly double the annual average amount.

The freezing rain began falling on January 5 and didn't stop until January 10. The usual casualties—trees, power lines, and airports—suffered losses. What was so extreme about this storm, though, was that more than 1.5 million people in eastern Ontario alone were without power and another 3 million were without electricity in Quebec.

The relenting rain and cold weather would prove fatal for those without alternative heating sources, and shelters had to be opened up to accommodate them. A staggering 28 people died during the storm, many from hypothermia, while 945 were injured, many due to carbon monoxide poisoning from the improper use of generators.

All told, 130 power transmission poles had fallen, as well as an estimated 30,000 utility poles.

A total of 57 Ontario communities were declared disaster areas and 200 in Quebec.

> **Storm Stats**
>
> Estimates after the Canadian ice storm were that it would take as much as 40 years before the maple syrup production in Quebec could return to normal because of the ice's damaging effect on the maple trees.

Military personnel were called in to canvass door-to-door to make sure people were okay. They also cleared storm debris and provided medical help to those in need.

The storm wasn't only confined to Canada; Maine was declared a disaster area, as well, and the area from the New England states to New York suffered losses of life and property. National Guardsmen were sent into many states, including New York.

## The Least You Need to Know

- Freezing rain differs from sleet in its composition and genesis.
- Ice jams on rivers can be devastating.
- The southern United States has seen significant ice storms.
- More than 4 million people were without power during the Canadian ice storm of 1998.

# Chapter 12

# The Deep Freeze

## In This Chapter

- Cold fronts
- How cold can you go?
- Cool climates
- Keeping warm

Ever "Eskimo kiss"? The rubbing of noses, rather than smacking of lips, isn't just a cultural thing; it's for practical purposes. Lips, which usually are wet, tend to stick together when you kiss someone in a climate that's often several degrees below zero.

In this chapter, we'll look at where and how cold weather originates, some extremely cold places, and how to avoid losing an appendage—or your lips—to the cold.

## Where Extreme Cold Weather Comes From

The earth tilts on its axis as it rotates around the sun. The farther from the sun, the colder it is. For certain areas, such as the North Pole, it's cold a whole lot. For others, Cancun, Mexico, for instance, it's warm most of the time.

Seasonal climates get colder when those areas are farther from that great heating source.

## Putting Up a Front

So it's a given that the Yukon Territory of Canada is going to be really cold for most of the year. But what about Nashville, Tennessee; Washington, D.C.; even Atlanta, Georgia? Surely, these are relatively decent climates, not super-near, or super-far from the sun.

Air moves. And when it moves significantly, we call it a front.

And there are cold and warm fronts.

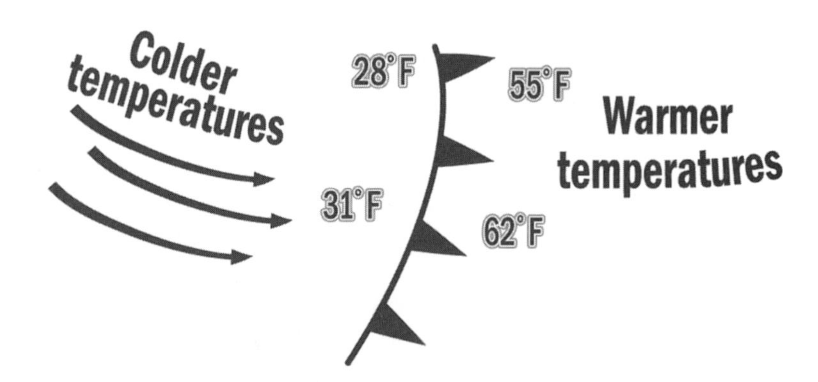


*Cold front.*

*(Courtesy of the National Weather Service.)*

A cold front happens when a cold air mass replaces a warmer one. These fronts generally move in a northwest to southeast direction. When that cold front sweeps through a region, the air behind the front is drier and colder than the air ahead of it. A 15-degree drop in the first hour isn't at all odd.

When watching the weather on the TV news or checking out the weather page in your newspaper, that curved blue line with all the triangular points protruding eastward represents the cold front.

**Eye of the Storm**

If you're ever at the South Pole on a severe below-zero day, try dumping out your hot coffee. By the time it hits the ground, it will be frozen.

Cold fronts are usually ushered in by a pressure change, as well as a noticeable change in weather—rain, snow, wind—sometimes very severe. All of these are characteristic of most cold fronts moving through. In fact, on the east side of a cold front the temperature could be 60 degrees, while following behind it—just a short distance away—it could be 45 degrees.

So, when that cold Canadian air catches a ride on the rails of a cold front, folks in Atlanta could awake to frost on their windshields.

## Wind Chill

Folks in Pelly Bay, in the Northwest Territory of Canada, are used to the cold. Below-zero temperatures aren't uncommon throughout the winter months. But on January 28, 1989, the temperature dropped to 59 degrees below zero Fahrenheit. Even worse, the wind chill on that day pushed the dangerous temperature to minus 131!

The unbelievable cold set the wind chill record for Canada.

**Actual temperature** (in degrees Fahrenheit)

| Wind speed | 50 | 40 | 30 | 20 | 10 | 0 | -10 | -20 | -30 | -40 | -50 |
|---|---|---|---|---|---|---|---|---|---|---|---|
| | **Equivalent temperature** (factors in wind speed) | | | | | | | | | | |
| 0 | 50 | 40 | 30 | 20 | 10 | 0 | -10 | -20 | -30 | -40 | -50 |
| 5 | 48 | 37 | 27 | 16 | 6 | -5 | -15 | -26 | -36 | -47 | -57 |
| 10 | 40 | 28 | 16 | 4 | -9 | -21 | -33 | -46 | -58 | -70 | -83 |
| 15 | 36 | 22 | 9 | -5 | -18 | -36 | -45 | -58 | -72 | -85 | -99 |
| 20 | 32 | 18 | 4 | -10 | -25 | -39 | -53 | -67 | -82 | -96 | -110 |
| 25 | 30 | 16 | 0 | -15 | -29 | -44 | -59 | -74 | -88 | -104 | -118 |
| 30 | 28 | 13 | -2 | -18 | -33 | -48 | -63 | -79 | -94 | -109 | -125 |
| 35 | 27 | 11 | -4 | -20 | -35 | -49 | -67 | -83 | -98 | -113 | -129 |
| 40 | 26 | 10 | -6 | -21 | -37 | -53 | -69 | -85 | -100 | -116 | -132 |

Winds speeds greater than 40 mph have little additional effect.

Little danger for properly clothed person | Increasing danger | Great danger

*Wind chill.*

*(Courtesy of the National Weather Service.)*

Add wind to cold and … you get the picture. The greater the wind, the greater the cold feels. Exposure to these subzero temperatures can be deadly (more on that in a moment). Wind chill isn't recorded regularly, but examples, like Pelly Bay's, are on some pretty firm ground, so to speak.

**Inside the Storm**

Many factors contribute to cold weather, but wind often is a major component to extreme cold.

At the Antarctic U.S. South Pole weather station, minus 81 degrees Fahrenheit is the average low for July, August, and September. Looking at the wind chill chart above, add a 10-mile-per-hour wind to that negative 81 degrees and now we're talking an equivalent temperature of minus 116, and so on.

# Can It Get Any Colder?

For people used to the extreme cold, below-zero temperatures are a way of life. Many folks living in Canada, the northern United States, and Alaska have made the harsh winters more tolerable with solar heat, efficient woodstoves, heated floors, electric engine blankets, plug-in heaters, super-insulated homes, and remote car starters.

But when temperatures plummet into the negative 30s, 40s, and even 50s, survival becomes the rule of the day.

## Chilling Effect

Ice in the Gulf of Mexico? It happened. In February of 1899, North America was gripped by a relentless cold front that sent temperatures tailspinning below zero in many states that had never even known it could become that cold.

Washington, D.C., saw negative 15 degrees on February 11; Tallahassee, Florida, hit 2 degrees below zero on February 13; Montgomery, Alabama, saw minus 12 degrees; and in Logan, Montana, residents were stung with minus 61-degree temperatures.

**Storm Stats**

The year 1899 brought cold weather records to Ohio, Louisiana, Nebraska, Florida, and Washington, D.C.

Single-day cold records in 1899 included Pittsburgh, Pennsylvania, at minus 20 degrees on February 10; Cleveland, Ohio, with negative 16 on the same day; Erie, Pennsylvania, at minus 12, also on February 10. At Charlotte, North Carolina, temperatures fell to minus 5 degrees on February 14.

Ice jams formed on the Ohio, James, and Tennessee rivers. The Mississippi brought ice down to New Orleans, Louisiana, and on February 19, that ice was seen in the Gulf of Mexico.

On February 11, 1899, a high-pressure area just north of Montana was replaced by an extreme *Arctic High* that stretched from Maine to Texas and across the Rockies.

**def•i•ni•tion**

An **Arctic High** is an extremely cold high-pressure system originating over the Arctic Ocean.

By February 12, Camp Clark, Nebraska, recorded 47 degrees below zero. A day later, the Arctic high moved through the Gulf states and a low-pressure system brought major snow into the mid-Atlantic and points north in New England.

On February 14, the East Coast blizzard was followed by the Arctic High and the Carolinas saw subzero temperatures as the storm moved through.

The weather event became known as the Great Cold Wave of February 1899. The loss of lives was estimated at about 100 people from January 29 to February 13, also due to snow and avalanches.

## Warm Places Gone Cold

President Jimmy Carter declared 25 Florida counties disaster areas in January 1977. It wasn't a hurricane, a flood, or even a tornado. It was bitter cold that decimated the citrus and vegetable crops and brought snow to parts of the United States that had never seen the fluffy white powder.

The cause of the storm was a cold front that pushed through the country, farther south than normal, in mid-January. By the nineteenth of January, below-freezing temperatures had been recorded throughout the state. Naples logged 26 degrees and in West Palm Beach it was 27 degrees. Miami Beach even saw a temperature of 32 degrees and Dade and Broward Counties saw temperatures as low as 20 degrees. Temperatures there remained below freezing for 14 hours.

**Storm Stats**

Florida's economy suffered a major loss in the winter of 1977. Dade County's loss in citrus and vegetables was estimated at $100 million.

Highways were covered with ice during the weather event and hundreds of accidents were caused by the slippery conditions. At sea, boats were damaged when high winds offshore pummeled the coast.

Agriculture suffered great losses. Frost-protection systems couldn't keep up with the severe cold, and tangelos, tangerines, and temple oranges were almost completely destroyed.

# People Really Live Here?

If you live in Snag, Yukon Territory, you're probably used to the jabs from folks who find it to be a funny name. Snag took its name from the hidden stumps and outcrops in the White River tributaries, where boats would often become snagged. But you're living in Canada's Yukon, so chances are there aren't a whole lot of people around you.

But just north of a remote military station near the Alaskan border, the few people living there suffered some remarkably cold weather on February 3, 1947. In fact, it set the record for the coldest temperature in North American history: 81 degrees below zero.

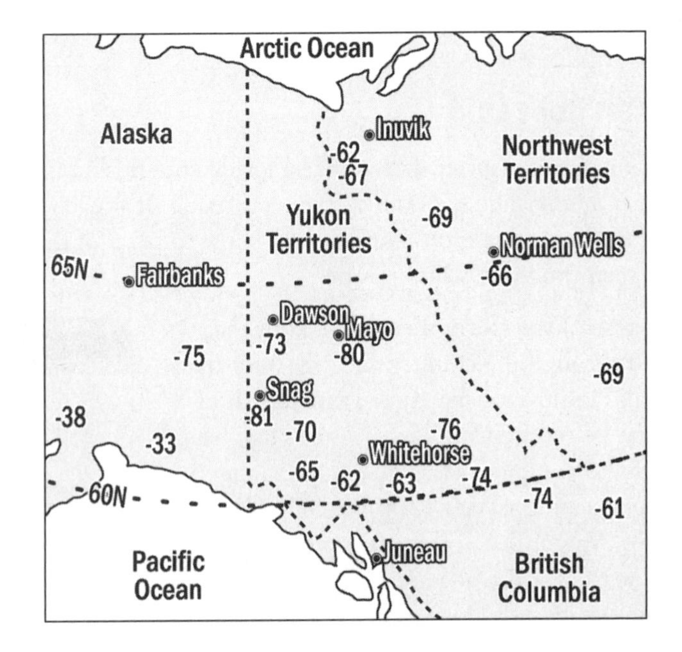


*Snag, Yukon.*

*(Courtesy of the National Weather Service.)*

Snag is tucked in the southwestern corner of the Yukon and had been the site of a World War II–era auxiliary military airfield about 5 miles south of a village that had a population of about 10 in 1947. The airfield was part of a weather station and the base was used by about 15 men as an emergency landing strip for both Canadian and American military planes. The village was a trading post, which no longer exists.

> **Storm Stats**
>
> At minus 81 degrees, staff at an auxiliary military airstrip in Snag reported that they could clearly hear voices from miles away. The nearest village was 5 miles away. It stands to reason: sound travels farther over frozen surfaces.

Snag is a hard place to live in. The day before the record-breaking temperature, the thermometer registered minus 80—just a degree shy of the historic mark. Typically, the winter daily high temperatures average a balmy 13 degrees below zero with lows averaging about minus 32 degrees.

Temperatures reaching 50 below are not uncommon, and the area's geography has a lot to do with it. Mountain ranges nearer to the coast tend to block Pacific air from coming into the region, so Snag is largely exposed to the cold Arctic air.

A good deal of the frigid air that plagues Snag makes its way down into the United States on particularly cold winter days.

Cold? Blame Snag.

To get to 81 degrees below zero takes a little bit of work. The geography is definitely a key ingredient, but the winter of 1946–1947 had some other parts to it that threw the cold recipe a twist.

**Storm Stats**

Colder temperatures have been officially recorded in Antarctica, Siberia, and Greenland, but minus 81 degrees Fahrenheit is the official record for North America.

Arctic air had been trapped by a strong westerly circulation upper-atmosphere pattern for much of the winter over Alaska and northwest Canada in sort of a cold dome. Clear skies and calm winds allowed the cold air to spill into the low areas (remember, cold air falls) and overnight temperatures were plummeting to 50 below.

It's important to note here that a typical thermometer contains mercury. Well, at minus 39 degrees, mercury freezes. So, if you want to measure really cold weather, alcohol-based thermometers must be used. The very last mark on that alcohol thermometer is minus 80.

**Storm Stats**

The Canadian Weather Service took three months to test and examine the thermometer used to record the coldest North American temperature in history.

The only way to measure temperatures lower than that was to estimate with a set of dividers to determine the slide's position. The estimation was minus 83 degrees. The thermometer was sent to the Canadian Weather Service in Toronto, rigorously examined, and eventually confirmed at 81.4 degrees below zero.

## Staying Warm in Extreme Cold

Cold can be a killer. Exposure to extreme cold temperatures and wind or cold weather and moisture for an extended period of time can adversely affect one's heath and well-being. Two of the most common winter ailments are frostbite and hypothermia.

Old Man Winter provides the venue for great days of skiing, skating, and sledding, but also a dangerous one for losing a finger or a life.

More than 700 people die each year in the United States alone from hypothermia, and thousands of people become susceptible to frostbite.

## Frostbite

Frostbite sneaks up on a person. Usually, it's not until someone else notices that your ears have gone glossy and white or until you come indoors and feel a searing sting that you know something is up.

There are degrees of frostbite, and it's far more common than you may think. On the superficial end of the spectrum, frostbite looks white and waxy, and sometimes the affected skin even takes on a grayish-yellow patch. The skin will feel numb and cold, but not frozen through. At this point, it's important for you to quickly get indoors, where it's warm. Take off clothing that might impede circulation and get medical attention. Be sure fingers and toes don't stick together and be sure there's no moisture between each of them. A sterile, dry piece of gauze will do the trick. Elevate the affected area to reduce swelling and pain. If the most immediate medical attention will be more than an hour away, immerse the frostbitten area in warm water, about 102 to 106 degrees Fahrenheit. Be sure that it's not too hot.

Deeper frostbite will need immediate medical attention.

**Storm Stats**

Walking on frostbitten feet or rubbing frost-bitten fingers can actually damage them more.

Prevention is the best method to avoid frostbite. Wear warm, layered clothing and cover up extremities—fingers, toes, ears, nose, chin. Stay dry, and try not to sweat, either. Eat a well-balanced diet, avoid alcohol and caffeine, but drink fluids—especially warm ones—to hydrate.

## Hypothermia

When the temperature in the body falls below 95 degrees Fahrenheit, it's called hypothermia.

People with poor circulation, such as the elderly, especially are susceptible to cold. Babies, too, can become affected, and those who spend long periods of time outdoors—hunters, hikers, skiers, and homeless—are the most common victims.

Hypothermia is very dangerous and can cause severe injury or death. The condition can bring a change in mental status including disorientation and a lack of reasoning, as well as a cool abdomen, uncontrollable shivering, rigid muscles, irregular heart beat and respiratory rates, and even unconsciousness. People can suffer memory loss, slurred speech, and/or drowsiness—all signs of hypothermia. Low energy in babies and bright red, cold skin can be common signs.

Exposure to cold and wet conditions are the main cause of hypothermia.

Medical attention is needed immediately. Get the victim out of the cold and warm him or her by adding blankets, pillows, towels, or newspapers around the victim. Cover his or her head, too. Be sure the victim is dry.

Also, handle him or her gently, as sudden or rough movement can cause cardiac arrest. Keep the victim horizontal.

**Storm Stats**

Folks stricken with hypothermia may not know it—not because they are too cold to realize it, but because the condition causes delusion and lack of reasoning.

## The Least You Need to Know

- Cold fronts from cold climates disburse cold weather.
- February 1899 was one of the coldest years on record in North America.
- The village of Snag in the Yukon set the North American record for lowest temperature at minus 81 degrees Fahrenheit on February 3, 1947.
- Frostbite and hypothermia are the two leading causes of cold-related injuries and death.

# Part 4

# Keeping Your Head Above Water

More people are killed by water than any other type of storm. We're talking about rain and flooding here, and you can bet that thousands of people have lost their lives, and hundreds of thousands more have lost their homes, when the rain came hard and heavy. But let's not forget about breaking dams, levee breaches, and seawalls buckling under massive amounts of rainfall and snowmelt. Whole floodplains have been reclaimed when freak storms came calling.

In this part of the book, we'll look at severe precipitation, and how to keep your head above water.

# Chapter 13

# 40 Days ... Rain, Floods, and Landslides

## In This Chapter

- Extreme rain
- Mitigating circumstances
- Rainy day people
- Slip-sliding away

Extreme rain means more than flooded basements and streets. Flash floods that can drown thousands, rake entire landscapes, and decimate livestock are among the most dangerous weather phenomena on the planet. And let's not forget mudslides. Whole chunks of earth breaking away as the rains force their grips loose, sending thousands of tons of mud down slopes in excess of 100 miles an hour. Entire villages can be claimed in a matter of seconds.

And it's not all Mother Nature's fault. People, as we will see, have created some of the problem. In this chapter, we'll look at what extreme rain can do.

# Rain, Rain, Go Away

If you live in Lloro, Colombia, you'd better have a raincoat. The South American country is home to the greatest average rainfall in the world. And people who live or pass through here know it all too well.

During a 29-year average, Lloro, tucked in the northwest corner of South America, receives an average of 523.6 inches of rain a year.

That's about 10 times more rain than fairly wet United States or European cities.

While rain in moderation can be rejuvenating, rain in excess can be deadly. With it come floods, mudslides, and disease. It can pull homes off their foundations and bury entire cities in an earthen tomb.

# Extreme Rain

Rain forms in the clouds where millions of tiny cloud droplets of melting ice merge to form a raindrop. One single raindrop has a volume more than a million times that of a cloud drop.

But what causes severe rain—rain that doesn't seem to stop, or a season that is filled with more rain than ever, or a perpetually rainy area like Lloro?

## Why All the Rain?

When it comes to rainfall, an area can experience several intensities from light rain showers to heavy thunderstorms to continuous rain and, finally, flooding. Let's focus on the continuous rain. Sometimes storms get "hung up" in an area. They plague the region with heavy rain for days. What causes this particular scenario? A couple of things can happen to give a region heavy rainfall. Many times in the upper levels of the atmosphere there are areas of low pressure that decide to do their own thing. They basically get cut off from the mainstream flow of the jet stream. These areas of low pressure are appropriately named "cut-off lows."

**Eye of the Storm**

The Johnstown Flood of 1889 in Pennsylvania stands as one of the nation's greatest tragedies. An estimated 2,200 people were killed when rain from a massive storm made river levels rise, ultimately breaching a dam.

*Jet streams* typically move west to east all around the world with little storm formation. However, a jet stream can quickly meander and produce a strong north to south flow. Colder air moves south and warmer air moves north. Strong winds drive colder air to the bottom of a jet stream in an area called the trough.

**def•i•ni•tion**

The **jet stream** is an eastbound wind current often exceeding 250 miles per hour at altitudes of 10 to 15 miles.

Oppositely, warmer air rides a ridge in the jet stream and advances north. Storms are created. The stronger southward-moving winds can cause the base of the trough to separate from the main flow and form a storm that gets cut off from the main jet stream flow. When this happens, regions can experience dismal, wet weather for days.

Another way that continuous rain can form is when the upper airflow runs parallel to a cold front. This can cause the cold front to stall in an area and produce heavy, continuous rain. It would take the upper airflow to change directions to finally kick the cold front out of the area and advance it to another region, putting an end to the rainfall.

That's all well and good, but doesn't it seem like some years are just plain rainier than others? Well, they are.

If you live in an area for a few years, you begin to learn when your rainy seasons occur. Sometimes, a particular season can have more rainfall than others. For instance, you can have average rainfall for your area in the summer for a couple years in a row. Then one year, you get storm after storm after storm. Your seasonal rainfall is well above average.

How does this happen? The position of the *polar jet stream* plays a big part in how much precipitation falls and how cold or warm your temperatures are.

If the polar jet stream situates itself north of your area, you experience warmer temps and drier conditions. Oppositely, if the polar jet stream moves south of your location, it will bring colder air and stormier weather your way. So that could easily mean more rainfall for your region during a season that is usually drier.

Also, *El Niño* patterns can cause an area to experience more rainfall in a season. For example, in El Niño years, the pressure changes in the Pacific Ocean.

**def•i•ni•tion**

The **polar jet stream** is a ribbon of air that flows from the west toward the east in the upper troposphere.

The natural and unpredictable results of clouds, storms, winds, ocean temperatures, and currents in the equatorial Pacific Ocean is called **El Niño.**

Low pressure forms off the coast of South America and high pressure forms near Australia, which causes warm water to travel eastward along the equator. This brings warmer water to the eastern Pacific Ocean and causes the overall weather pattern to change, not only in South America, but all over the globe. Winters are warmer in the northern United States. Hurricanes develop less in El Niño years, too. Wetter than normal conditions are experienced in the eastern United States Therefore, more rainfall would occur in an area.

**def•i•ni•tion**

**Orographic lift** happens when relatively warm air is pushed up a slope and cools. Any moisture is squeezed from the lift and thunderstorms can happen.

Some places are wetter than others simply because of their topography. Mountain chains play a big role in rainy weather versus dry weather. Basically, rainfall that occurs over a mountain is formed by approaching storms. But the rainfall is enhanced by *orographic lift*. On the west side of a mountain, air rises. The air is slowed by the mountain barrier. The air rises and condenses into clouds and precipitation.

Most of the moisture gets rung out and rain falls. Once the air moves to the eastern side of the mountain it is much drier. The air flows down the mountain and warms up, giving the area less cloud cover and less rainfall. Many deserts are formed on the eastern side of mountain chains. For example, the West Coast, California in particular, can get a lot of rainfall from the Pacific Ocean. However, once that air moves over the Sierra Nevada mountains, it dries out and warms up.

Arid regions are formed just east of the mountain chain. Some of the hottest temperatures and driest conditions in the United States occur in Death Valley, California, which isn't too far from the Pacific Ocean.

Another example of orographic lift occurs in the Appalachian Mountains. For instance, the weather in Pittsburgh is cloudier and wetter than that of cities just east of it such as Harrisburg and Philadelphia. Pittsburgh is situated just west of the Appalachian Mountains. Air rises up the west side of the mountain chain and condenses and forms rain. Once the air dries out and moves down the eastern side of the mountains it warms up. Temperatures in Harrisburg and Philadelphia are a few degrees warmer than in Pittsburgh.

The central and eastern Pennsylvania cities have less rain and less snow. Finally, in Washington State, the Cascades provide different climates depending on which side of the mountain chain you live. The western side of the Cascades gets hit with a lot more rain. Once you travel to the eastern side of the chain, the air dries out and much less rainfall occurs.

## When the Rain Comes

As we've seen, when it rains it pours. And it pours a lot in Lloro.

But other places see extreme rainfalls, too. For 38 years, Mawsynram, India, averaged 467.4 inches of rain a year, which makes this Asian region the second rainiest area on the globe. But Hawaii is home to boatloads of rain, too. Mount Waialeae in Kauai received 460 inches in a 30-year average. Debbundscha, Cameroon, in Africa also broke the 400-inch mark on a 32-year average, with 405 inches of rain. Rounding out our top five, we return to Colombia, where Quibdo received 354 inches of rain in a 16-year average.

That's a whole lot of rain. And sometimes it's deadly.

**Storm Stats**

The costliest single flood from rain in U.S. history happened throughout the Midwest in 1993, totaling $18 billion in damages.

Each year in the United States, an average of 140 people lose their lives in floods. It's the number-one storm killer, on average, in the country. More than $5 billion in damages occurs every year, and despite mitigation and education efforts, those figures grow every year.

Many reasons exist, and we'll talk about the human factors that cause or contribute to flooding later in this chapter, but more people are moving to waterfront property every year and more development is occurring in watershed areas and in floodplains that once soaked up the rain.

In fact, in the United States in 2000, about 3,800 municipalities with more than 2,500 people were located on floodplains.

**Eye of the Storm**

Ten people died in Conchita, California, on January 10, 2005, when a mudslide from a massive rainstorm triggered a landslide that instantly buried homes under several tons of mud.

A massive storm from late December 2004 to mid-January 2005 nationwide caused millions of dollars in damages and claimed 20 lives. Most people will remember the winter storms in California during that winter with flooding, great snowfalls in upper elevations, and the heart-wrenching mudslide in Conchita.

What is known as "The Pineapple Express" was the cause of it all. The colloquialism refers to a subtropical jet stream that ships moist air to the West Coast from the tropics and Hawaii. As much as 10 inches of rain was dumped on parts of California in just a few days' time, and the Sierra Nevadas saw almost a foot of snow.

*More than 3,000 homes were damaged by flooding and mudslides in southern West Virginia in 2001 when 8 inches of rain fell. One person was killed and several were injured.*

*(Courtesy of the National Oceanic and Atmospheric Administration [NOAA].)*

*Many towns in the Mississippi River Valley saw extreme flooding during the Great Flood of 1993.*

*(Courtesy of the USGS.)*

In all, Los Angeles witnessed a record 15 wet days from December 27, 2004, to January 10, 2005, with 16.97 inches of rain. That's the greatest 15-day rainfall average since 1877. More than 20 inches of rain fell in Beverly Hills, Santa Barbara, and Ventura between January 6 and 11.

By January 18, Las Vegas reached 96 percent of its seasonal rainfall average with 4.42 inches of rain.

Reno's North Hills saw 7 feet of snow, the Sierra Nevada ski resorts' totals were as much as 6 feet, and 10 feet of snow fell in Tahoe from December 28 to January 11.

The same storm brought an inch of rain to much of Mississippi as it moved eastward on January 13. The Ohio Valley saw 2 inches and rain even fell on the East Coast as the storm left the country.

**Storm Stats**

During the western winter storms of 2004–2005, even the desert saw great rainfall. The Mojave received 119 percent of its seasonal rainfall total with 5.15 inches.

In all, more than 20 people were killed in the storm. Landslides, floods, and car accidents were the most common venues for tragedy. Power outages, rescues, evacuations, and transportation delays and the closing of airports were common.

Total damage from the storm was estimated at more than $100 million.

# Human Factors

No doubt, the more humans build, the more Mother Nature takes away. Perhaps the tree-huggers in the 1960s and 1970s had part of the equation right—if we exploit the planet, it's going to turn around and bite us back.

Clear-cutting and poor forestry practices, greedy farming, the tearing up of wetlands, replacing absorbent soil with impervious surfaces ... We're doing the planet great harm and terra firma is fighting back.

That's not to say we deserve it. In so many cases, we make mistakes—but we do have to learn from them.

## Not Seeing the Forest for the Trees

Since poor forestry practices came under intense scrutiny in the late twentieth century in the United States and Canada, better management practices are being applied. However, more homes in flood-prone areas, such as the Southeast's floodplains, mean more cutting of forests. We'll look at the hazards of not mitigating wetlands in a bit, but many nations around the world are still light-years behind.

South American rainforests and Asian forests have seen vast reductions—once-lush ecosystems left barren and unprotected from the elements. Sun parches the delicate soil, rain washes the sediment away, and wind blows the dirt-turned sand, kicking up sandstorms and ripping away any semblance of a once-absorbent terrain.

*(Courtesy of the National Institute of Space Research/ Brazilian Government.)*

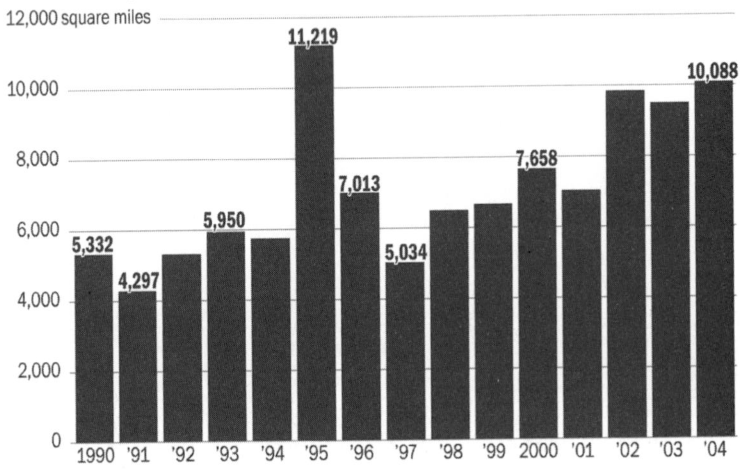


**Eye of the Storm**

The worst flood in Chinese history was in 1954 when 30,000 people were killed by a Yangtze River flood.

Trees have roots, and roots absorb water like a sponge. Trees drop leaves, and leaves lessen the impact of large raindrops on the soil.

It's all interrelated.

Now, when it rains, it floods, washing away the very nutrients and seed that once sustained the forest's life.

For the first time, the Chinese government in 1998 admitted that its forestry practices were degrading the environment and had become a factor in why the nation's floods were increasing in severity.

Unfortunately, the announcement had to follow one of the worst floods in China's history. August floods of the Yangtze River killed an estimated 2,000 people and affected more than 240 million people in 28 provinces. An amazing 13.8 million people were relocated and 5.5 million homes were destroyed.

The loss of trees was a huge part of 1998's Yangtze River flood. In the past few decades, the river's watershed has lost 85 percent of its forest cover.

## Farming Frustrations

Poor agricultural practices, like deforestation, can be equally detrimental to an ecosystem and even can help generate more storms.

In Chapter 8, we discussed the powerful dust storms in China and Mongolia. The lack of a penetrable ground cover can cause the ground to heat up and remain warm, which helps lift the dust into the atmosphere when a cold front passes. Some of the largest dust storms in history have occurred in China.

And when that sediment blows away, it takes with it vegetation that once held firm to the earth. When the seasonal rains come—and they come hard—flooding is more prevalent, if not deadly.

Poor farming practices, such as overclearing land, not planting on cleared land, and overgrazing are causes for degradation. The less penetrable the surface, the less absorption. And when that happens, the river's going to rise.

The Egyptian Nile River is a good case study. Before the Aswan Dam was built in 1970, summers would bring massive flooding to the banks. But these floods were seen as a blessing.

> **Eye of the Storm**
>
> The annual summer floods in Egypt once had been known as "The Blessing of the Nile."

When the river rose, it would bring thick, nutrient-packed mud onto the banks. The farmers immediately would plant their seeds and the crops would flourish without fertilizer.

But as the Egyptian population increased, so did the need for more food, and the answer was the dam.

The dam would stop the annual floods, but allow the river waters to be held in a reservoir and released throughout the year as needed. The result was year-round farming.

The problem is that 98 percent of the river's sediments are trapped in that reservoir and massive amounts of fertilizer are needed. Add to that equation that the Nile Delta is no longer being nourished with river sediment, meaning no more dirt is being added to the banks downriver, and without those natural reinforcements, erosion and high levels of salinity are taking their toll.

## Wetland Woes

Areas around the world are discovering that it doesn't pay to build on floodplains. Still, the people come. They want homes close to the ocean or rivers, but find themselves drying out after the flood.

Blame it on growth in these areas. Population booms have occurred throughout the Southeast, from the Gulf Coast to Florida and into the Carolinas and Virginia. According to 2004 census data, the population of Florida has tripled since 1960, with 17 million people. Between 2000 and 2004, 29 of the 50 fastest-growing counties in America were on the East Coast and Gulf Coast. In the 10-year period from 1990 to 2000, the Sunshine State saw a 23 percent population increase.

**Storm Stats**

Each day, about 1,000 people move to the state of Florida.

The southeastern United States, including the Carolinas on down to Florida as well as the Gulf Coast from Texas into Alabama, has seen tremendous floods, and the water will keep rising if the loss of wetland habitat isn't mitigated properly.

In the United States, approximately 100 million acres of wetlands remains of the original 215 million acres that existed in the continental United States 200 years ago. California, Indiana, Illinois, Iowa, Kentucky, Missouri, and Ohio have lost more than 80 percent of their wetlands in that time and 22 states have lost at least half.

Of course, agriculture in many areas is the reason for this loss. From the 1950s to the 1970s, an average of about 458,000 acres were lost each year. The federal Clean Water Act in 1972 has helped—now the United States averages a loss of about 117,000 acres each year on average. That's still too much. Just do the math: there are 100 million acres left—roughly the size of the state of California—and we're losing just over 117,000 acres a year. At this rate, every decade, the United States would lose more than 1 million wetland acres. In just over 850 years, the wetlands would be gone.

Florida is one of the fastest-growing states in the country, with an average of 1,000 people moving there each day. Of the 100 million acres of wetlands remaining, Florida has the most with 11 million.

In China, the improper use of land has led to degraded fields: 34.5 percent of degradation has been caused by overgrazing, 29.5 percent by the destruction of forests, 28.1 percent by the improper use of agricultural land, and 7.95 percent by the improper use of water resources, according to a United Nations environmental report.

With nearly 1.3 billion people estimated to live in China now and nearly 1.5 billion by 2025, environmental safeguards are needed.

Wetland mitigation processes are needed to ensure that these floodplains are protected. Parking lots, rooftops, and driveways, for instance, are impermeable surfaces—they don't absorb water. Instead, they allow storm water to run off into the rivers and marshes, taking with it the substances needed to filter pollutants and keep vegetation growing. Many folks move to the waterfront to get a good view of the water. But chopping down trees and native brush, as well as trying to maintain lush green lawns with chemicals and overwatering, cause dangerous erosion.

Building retention ponds, ensuring that compromised wetlands are traded for new watershed areas—so-called wetland banking—and not allowing the removal of native growth of vegetation are some of the tools governments can use to offset destructive practices.

In Beaufort County, South Carolina, the local government charges residents a fee to ensure that a funding pool exists to mitigate wetlands and storm-water runoff.

Is it enough? Probably not. But it's certainly a step in the right direction.

## Flood Warning

In the United States, "The River" is the Mississippi. It's one of the biggest rivers in the world and depending on the year—the river tends to reinvent itself—it's roughly 2,300 miles long. Combined with the Missouri River, it is one of the largest watersheds in the world at about 2 million square miles.

*Grafton, Illinois, took an extremely hard hit when 1993 flooding rose to the rooftops.*

*(Courtesy of the USGS.)*

So when the mighty Mississippi floods—and boy, does it!—it has the potential to become a major natural disaster.

Such was the case in 1993 during the great Midwest flood. Damages totaled a staggering $15 billion and an estimated 52 people lost their lives. The levees that had been in place to thwart a disaster such as this were quickly overwhelmed.

The flood ranks as one of the worst natural disasters in the world and the largest ever to hit the United States. Approximately 15 million acres of farmland were destroyed by the flood, 75 towns were completely underwater, and 50,000 homes were damaged. For two months, all river commerce ceased, bridges were washed out, and roads were ripped up.

*An aerial view of the Missouri River after the great flood on June 30, 1993, shows the area of the Jefferson City Memorial Airport underwater.*

*(Courtesy of the USGS.)*

The floods began after as much as 48 inches of rain fell from June through August throughout the Midwest in the Dakotas, Illinois, Indiana, Kansas, Missouri, Iowa, Minnesota, and Nebraska. The rainfall totals were as much as 300 percent higher than average for some of these areas.

The soil saturated, the river began rising, and rising, and rising some more. A new record high stage of 47 feet was seen in St. Louis on July 20; in Kansas City a week later, the Missouri River crested at 48.9 feet. By August 1, St. Louis measured 49.47 feet and the Great Flood of 1993 was born.

Nine states over 400,000 square miles were affected and more than 1,000 levees were breached.

Despite our best efforts, sometimes there's no match for Mother Nature.

*The Missouri River flood levels at St. Charles, Missouri, practically buried the railroad station in 1993.*

*(Courtesy of the USGS.)*

# Extreme Mudslides

The tiny oceanside town of La Conchita, California, just outside of Ventura, was a sleepy beachside community on one of the most picturesque places along northern California's coast.

Along the highway are curiously posed homes, modest in build, steep in price, with vistas looking over the Pacific—a surfer's paradise.

But paradise turned to panic on January 10, 2005, when the earthen wall that nestled its residents between it and the ocean gave out. And within seconds, millions of tons of mud covered the place where the homes had been and people had lived.

Ten people were killed in the tragedy. It wasn't the worst mudslide in American history—not even in California history—but the tragedy hit home for millions of Americans who watched the story unfold on TV and read about it in newspapers for weeks, as husbands, fathers, neighbors, and children dug around the clock to get to their loved ones.

For two days, rescue crews and family members dug through the mud where 13 homes once stood. Another 23 homes had been damaged.

*A deadly mudslide near La Conchita, California, on January 10 claimed 10 lives and destroyed 13 homes within seconds.*

*(Courtesy of the National Oceanic and Atmospheric Administration [NOAA].)*

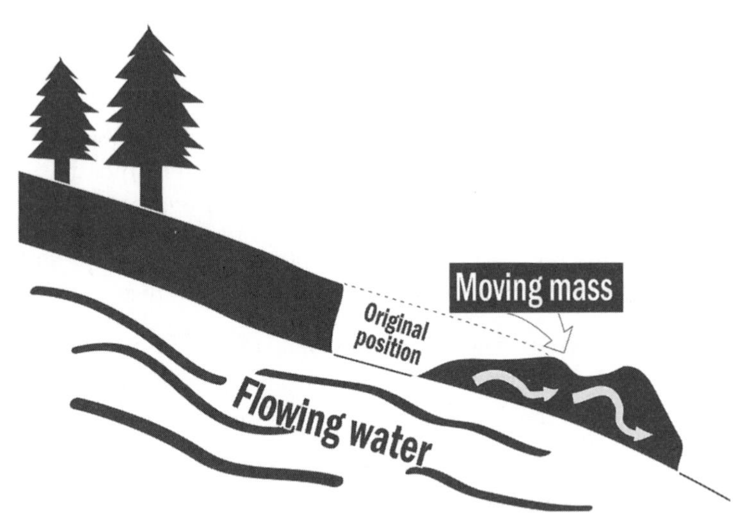


*How a mudslide occurs.*

*(Courtesy of the U.S. Geological Survey.)*

The La Conchita mudslide wasn't unlike most. Rainfall from a strong storm in a short period of time can cause the earth to break away, moving at a force and speed that is often lethal.

Each year in the United States alone, an average of $2 million in damages is caused by landslides.

California may be one of the nation's worst areas for mudslides, but they pale compared with those in Asia, India, South and Central America, and the Middle East.

In Nepal in July 2001, an estimated 100 people lost their lives after heavy monsoon rains caused a mudslide.

**Inside the Storm**

San Francisco saw more than 1,800 landslides overnight in January, 1982. A staggering 25 people lost their lives and property damage was reported at more than $66 million.

In Pakistan, also in July 2001, as many as 200 people died in mudslides triggered by 20 inches of rain in 10 hours. Sixty-five more Pakistanis were killed just days later when 40 houses were swept away by mudslides in Manshera.

In South Korea, also in the month of July, 40 people were killed when more than a foot of rain fell at a rate of almost 4 inches an hour and triggered a landslide near Seoul.

Historically, however, landslide fatalities in the thousands have occurred, the worst being December 16, 1920, in Kansu, China. More than 180,000 residents were killed in a landslide. In 1949, 12,000 people were buried in Khait, Tadzhikistan. In the Chiavenna Valley of Italy in 1618 more than 2,200 people died. In Rio de Janerio, Brazil, in 1966, 550 people died.

**Inside the Storm**

A mudslide can move at speeds in excess of 100 miles per hour.

## The Least You Need to Know

- Torrential rain and floods kill thousands each year worldwide.
- Humans play an increased role in causing flooding—more than ever before.
- The Great Flood of 1993 was the worst in American history.
- Mudslides can travel at more than 100 miles per hour.

# Chapter 14

# When the Levee Breaks

## In This Chapter

- Dealing with the deluge
- City underwater
- Mitigating circumstances
- After the flood

Floods kill more people each year on average around the world than any other type of weather-related event. Since 1900, flooding has killed more than 10,000 people in the United States alone. Add to that the thousands in Asia and other parts of the world and the picture becomes clear.

Floods kill.

They also are expensive. Property damages in the United States average $1 billion a year.

And floods are practically unstoppable. Despite building levees, channels, dams, and dikes, when the rains and storms come they can quickly overwhelm even the best antiflood measures. In this chapter, we'll look at the damage that heavy rains do and how humans keep fighting this losing battle.

# How Rainy Cities Cope with Floods

It's not all death and destruction. In fact, many cities in the United States and around the globe have semi-successfully figured out a way to mitigate Mother Nature's watery wrath.

Of course, the more humans build on floodplains the more at risk people become.

> **Storm Stats**
>
> After massive flooding in 1927, the longest series of levees in the world were built along the Mississippi.

In Chapter 13, we saw the damage that snowmelt, rain, and saturated soil can do when it's all in one place. The mighty Mississippi River was the example when in 1993 a massive storm caused 52 deaths and billions of dollars in damages. Fact is, 534 counties in nine states were declared federal disaster areas and more than 168,000 applied for federal assistance.

That flooding cost the federal government a staggering $4.2 billion, with $271 million in federal flood insurance and more than $621 million in federal loans, to get communities back on solid ground.

The Mississippi floods during the summer of 1993 were a great example of practically everything that could go wrong during a natural disaster. Stretching 2,300 miles, the Mississippi joins the Missouri River to form a 2 million-square-mile watershed that ranks among the largest in the world.

*Even farm country is no match for the flooding.*

*(Courtesy of the National Weather Service.)*

A series of *levees* had been built to hold back the water should the river rise. And for years those levees did their jobs. But when as much as 48 inches of rain fell from June to August, 1,000 levees were breached when the river rose to nearly 50 feet. Seventy-five towns along the banks were submerged, 50,000 homes destroyed, and 15 million acres of farmland ruined.

**def•i•ni•tion**

A **levee** is a human-made embankment that runs parallel to the river and is designed to prevent the river's floodwaters from overflowing.

So what's the solution?

First off, as we have mentioned, people are building on wetlands and in watersheds. The more we trade away that absorbent earth for asphalt and concrete, the more flooding we will have. To thwart that flooding so we can live on or near the water, we build dams that create lakes and regulate water flow through rivers. We also build levees, which are like dikes that run parallel to the water's flow, to hold back the rivers when they flood. Locks along rivers also even the flow of water.

**Eye of the Storm**

One thousand levees buckled when nearly 50 inches of rain caused massive flooding along the Mississippi River in 1993.

Thing is, Mother Nature had the answer all along in the floodplains. See, those flat, porous fields have taken care of all that river spill since the river first cut through the Midwest. There was no need for levees along the Mississippi. So can we really channel that water through without the floodplains?

Well, not in 1993 anyway. That water way, way up the river that should have breached the banks and percolated into the fertile soil instead became channeled down the river, where it grew higher and higher. Eventually, it was too much, and the levees gave up the ghost. And the water again reclaimed the floodplains.

## Modern-Day Atlantis

When Category 4 Hurricane Katrina came ashore on the Gulf Coast in 2005, with it came a powerful storm surge. It's not just the rain that causes massive flooding—storm surges can be catastrophic, as was the case in New Orleans.

Katrina claimed the lives of more than 1,400 people, and many of them drowned.

*FEMA's Urban Search and Rescue Task Force helps people evacuate a flood-ravaged New Orleans on August 31, 2005.*

*(Courtesy of the Jocelyn Augustino/FEMA.)*

The coastal storm surge from Katrina rose as much as 30 feet higher than normal tide.

The surge actually spread from Santa Rosa Sound in Florida straight up through Mississippi into Louisiana. Flooding was widespread miles up the Mississippi River.

Perhaps the worst part of the storm surge was the false sense of security in New Orleans's series of levees. When the surge rose, New Orleans, which sits about 6 feet below sea level, was barely clinging to life as the levees held. But within hours of the storm, with the storm surge still forcing the waters to rise, the levees gave out. Add to the equation that New Orleans is literally surrounded by water: Lake Pontchartrain, the Mississippi River, and the Gulf of Mexico. There was nowhere else the water could go.

**Eye of the Storm**

New Orleans sits 6 feet below sea level and is surrounded by the Mississippi, the Gulf of Mexico, and Lake Pontchartrain.

Despite one of the largest and most elaborate levee systems in the world—covering 350 miles—the pressure of the wall of water that was hammering down on New Orleans finally broke through, flooding the city and causing more than $70 billion in damages.

New Orleans's proximity to so much water, the Mississippi notwithstanding, poses great concern. The floods resulting from Katrina weren't the first time folks compared the Big Easy to Atlantis.

*Residents are evacuated from their New Orleans homes during Hurricane Katrina in 2005.*

*(Courtesy of Jocelyn Augustino/FEMA.)*

In 2001, two University of New Orleans geologists found that 40 percent of all the coastal wetlands in the United States were in Louisiana, but the Mississippi River Delta had lost 1,000 square miles of land from 1930 to 1990. The rate of wetlands loss was as fast as 25 miles a year in recent times. Most alarming, however, was that New Orleans has been sinking 3 feet per century, which may not seem fast, but it's eight times faster than the worldwide rate. Add the fact that the city is several feet below sea level, and Atlantis may not have been such an odd comparison.

Other cities' levees have fared better than New Orleans's, but the force of Katrina has left government leaders around the country wondering what would happen if a storm of that magnitude directly hit their cities.

The California Delta is a good example. A levee system that runs from northern California to Southern California carries freshwater as low as 20 feet below sea level. The protective levee network stretches some 1,100 miles over the Golden State. However, if one breaches, it fills up and sucks in water from the ocean; the system may not fare better, in the end, than not having a levee at all.

# Flash Floods

Generally speaking, floods build up over a period of time. In upstate New York, folks watch the ice break up on the river and melt, raising the water levels a few feet, but it's usually just some minor spillover until baseball season's opening day.

**def•i•ni•tion**

A **flash flood** usually happens when storms drop a lot of rain in a short time and begin major flooding within a few minutes.

But in Florida, a hard afternoon summer storm can dump a lot of rain in just a short amount of time, and a *flash flood* can reach full potential in just a few minutes.

Flash floods can be dangerous. Often, drivers will find themselves caught in a flash flood. You make a trip to the grocery store during a storm and by the time you check out, the streets are flooded—and still rising. Flash flood waters have been known to rise as much as 20 feet in just a few minutes. In that kind of rush, you don't want to be anywhere near it.

*The 1972 flood in Rapid City, South Dakota, destroyed 5,000 automobiles, including these cars that were parked at a dealership on East Boulevard.*

*(Courtesy of the USGS.)*

Imagine the debris that is carried through the force of the water rushing. Imagine not knowing where the deep spots are—especially when manhole covers are blowing off from the force of the rain. Two feet of water can sweep your car right off its tires.

It's not a good place to be.

A very bad place to have been on June 9, 1972, was in the Black Hills near Rapid City, South Dakota. A strong thunderstorm camped out near the Black Hills, dropping about 15 inches of rain over 6 hours. That rain swelled the Cheyenne River over its banks, as it did Canyon Lake and several tributaries, and within moments the floodwaters were making their way through Rapid City. Once the flooding started, there wasn't time to get to higher ground, and 238 people drowned.

*The Canyon Lake Dam failed during the flood of June 9, 1972, sending a surge of water into Rapid City.*

*(Courtesy of the USGS.)*

Approximately 750 acres of land near Rapid Creek were flooded; property damage totaled $164 million at the time. An estimated 1,335 homes were damaged in the flood.

It wasn't the first time Rapid City had seen disastrous flash flooding. As long as flooding has been recorded, Rapid City has seen its share. In 1878, a flash flood washed a freight train and 10 ox wagons downstream when the river rose 10 feet. Four people were killed in 1883 when Rapid River rose after days of rain. In 1890, 11 members of a wagon train were drowned when flash flooding occurred on Beaver Creek. In fact, the Rapid City area saw floods 33 times in the twentieth century, and many of them claimed lives, took out bridges, and caused millions of dollars in damages.

# Keeping High and Dry

With the exception of fire, floods are the most common natural disasters on the planet. And there aren't a whole lot of places that don't experience them.

Most areas can sustain some flooding, unless there are colossal amounts of water raining down, being pushed from a storm surge, or stampeding down the river from a failed dam. It happens.

Dams don't last forever. But dam breaches are usually caused by structural damage, poor design, or neglect.

As we found in Chapter 13, the Johnstown Flood of 1889 in Pennsylvania was caused by a breached dam. It was the second time the dam breached in its lifetime and was the worst flood caused by a dam break in American history. An estimated 2,200 people died in the flood.

**Eye of the Storm**

Approximately 2,200 people died when the dam in Johnstown, Pennsylvania, breached in 1889.

Outside of a sudden dam burst, though, there are a host of ways to be smart about flooding. In areas like Rapid City, warning sirens have been set up that sound when there's danger of flooding.

## Know Your Area

Knowing the dangers of where you live can help keep you afloat. Obviously, if you live in the Mojave Desert, flooding probably won't do you in. But if you live in Cape Girardeau, Missouri, you'd better watch your back. The city is bordered by a massive floodwall made to hold back the Mississippi when it breaches its banks. Still, folks in this sleepy little town south of St. Louis know what the flood signs are.

But it doesn't take life on the Mississippi River to become an expert. The first thing to know is whether you are at risk of a flood: are you living on a floodplain and is it above or below the flood stage water level? Your local government can help you with that one.

Make a list of items that you may need should a flood hit: flashlights and extra batteries, battery-operated radio, first-aid kit, canned food and bottled water, manual can opener, cash and credit cards, warm clothes, and sturdy shoes.

If you're in a flood-prone area, you'll need building materials, like plastic sheeting, plywood, tools and nails, a saw, pry bars, shovels, and sandbags. Be sure you will be able to plug drains in your sinks, tubs, and showers, to prevent backflow into your house caused by extra water pressure.

*The riverfront of Cape Girardeau uses a large seawall to keep the water out. But during the Great Mississippi Flood of 1927, the 40-foot river stage overwhelmed the town.*

*(Courtesy of the National Weather Service.)*

It's important to know what your evacuation route is, if your town has shelters available, and whether your town has flood alert signals. An out-of-state contact is a good person to have, should you have to leave your home for a period of time.

If the flood hits and there is time, turn off the main power switch and close the main gas valve in your home. Fill your bathtub and emergency bottles with clean water. Move valuables to upper floors or higher elevations. Tie down lawn furniture and equipment or bring it indoors.

**Eye of the Storm**

During Hurricane Katrina, people had to cut holes in their roofs to escape the quickly rising water in their houses.

Don't ever attempt to drive through a flooded area. Most drowning deaths happen to people trying to escape by car. Two feet of rushing water can easily sweep away a vehicle. Get to higher ground.

**Storm Stats**

Cars can be swept away during floods in just 2 feet of water.

Remember, too, that power lines may snap and poles may collapse. Many people are electrocuted and die during floods because electric currents pass through water very easily. Beware of animals—especially snakes—which may use your home for shelter.

## After the Flood

Even after the floodwaters begin to recede, you're still not safe. Many buildings are susceptible to collapsing after being water-damaged. Also, gas lines may have ruptured, so remember not to create a spark and keep the electricity off until a professional electrician has inspected your system.

Local officials will let you know when you can use the water supply; until then, boil your water, use your freshwater storage, or melt ice cubes.

**Storm Stats**

Nearly 25 percent of flood insurance claims come from properties considered to be at low or moderate risk of flooding.

Another important consideration is whether you need flood insurance. It's important to note that your homeowners or business insurance policy will not cover floods; you will need supplemental insurance.

It's also important to note that flooding in the United States causes more than $2 billion in property damage each year. Federal insurance plans can be purchased through the National Flood Insurance Program.

## The Least You Need to Know

- Flooding is one of the most destructive natural disasters on the planet.
- Hurricanes can inundate a city with water, as can levee breaks.
- Flash flooding can cause a rapid rise in water in just minutes.
- Know how to protect yourself and your family from the dangers of flooding.

Chapter 15

# Extreme Weather Seasons: Monsoons

## In This Chapter

- Two sides to every monsoon
- When the rains come
- The Arizona Monsoon
- Cultural differences

Funny thing, these monsoons. Here in the Western world, we think monsoons are storms. Well, some of us. That's because we hear things like, "The monsoon in India has killed 1,500 people."

Well, the monsoon didn't so much kill the people. The weather during the monsoon did.

And it can be pretty severe, my friend. Rain, flooding, landslides, snakebites, crocodiles, disease … all are things that happen when an especially wicked monsoon season winds up.

Or at least one of them. There are two monsoon seasons.

And we have them here, too, in America.

# Two Sides to Every Monsoon

Just as there are four seasons in America—winter, spring, summer, and fall—there are two seasons in Asia and India that are so pronounced—well, they have their own names.

> **def•i•ni•tion**
>
> A **monsoon** is defined as any wind that reverses itself seasonally. It is formed by the back-and-forth cooling and heating of land and sea. There are wet and dry monsoons.

Say it with me: *monsoon.* In fact, the very word *monsoon* is derived from the Arabic word *mausim*, which means "season." Arab sailors coined the term.

The two monsoons are the wet, summer, or southwest monsoon—call it what you will—and the dry, winter, or northwest monsoon (ditto on the naming). It's not a summer-winter thing, but more about the type of air moving through during the monsoons. Obviously, the wet monsoon sees more precipitation and warmer temperatures, while the dry monsoon is cooler and more arid.

> **Eye of the Storm**
>
> The world record for the most rain during a monsoon was in 1860–1861 in Cherrapunji, India, when 1,042 inches fell in the monsoon season.

Here in the United States, we don't share the same monsoons, but we do witness our own smaller-scale changes. More on that in a moment.

What we're talking about is the low-latitude climate from West Africa to the western Pacific Ocean. Solar radiation and the different land- and water-surface temperatures play a major role in the monsoons.

Planet Earth heats up and cools down at different rates depending on how much solar radiation can be absorbed by the land at any given time. Land heats up more quickly than the ocean, but loses that heat at night or during rainy spells. It absorbs more heat but can't store it like the oceans can.

*Sea breeze.*

*(Courtesy of the National Oceanic and Atmospheric Administration.)*

But those temperatures spill into one another via the so-called sea breeze. As the land heats quicker than the water, the hot air rises and the cooler, moist air fills the gap over the land. The air temperatures are evened somewhat. The same thing happens at night, only in the opposite direction. As the land loses its heat, the cooler air over the land now blows toward the water and the temperatures even out.

**Inside the Storm**

The temperature difference between the land and sea can be as much as 68 degrees Fahrenheit.

On a larger scale, this is the gist of how the entire planet stays fairly uniform. Sure, it's hot in the desert and cold in the Arctic on the same day but there is a balance that must circulate throughout the world and those massive air streams, like our very own jet stream, are the vehicle.

So what does that have to do with monsoons? That energy balance is more pronounced in some areas. Areas affected by monsoons usually see a more stark contrast.

Let's take a closer look at the monsoons.

## Summer of Discontent

The wet monsoon is the more monstrous season. This monsoon blows in from the southwest from roughly May to September, and when we think of monsoons we think of this one: a season of torrential rain, flooding, landslides, and strong thunderstorms affecting Asia and Bangladesh.

In fact, during this time of the year it's not uncommon to see areas, especially in India toward the Himalayas, with upward of 400 inches of rain.

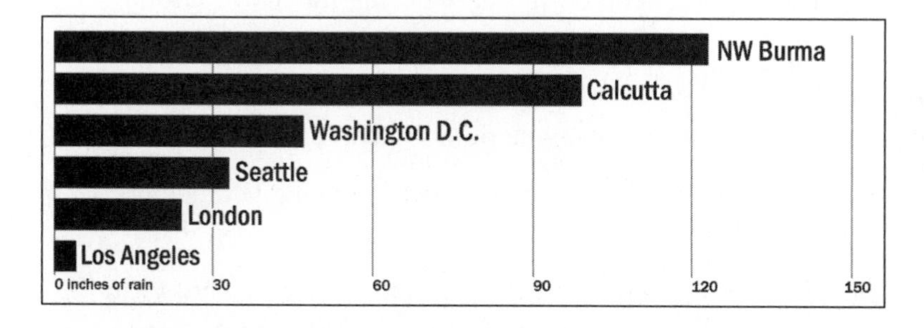


*Yearly monsoons.*

*(Courtesy of the University of Nebraska-Lincoln School of Natural Resources.)*

Temperatures heading into the summer monsoon in southwestern India generally reach 100 degrees. When that heat comes in the early season, it comes with the anticipation that cool air soon will follow from the ocean and the rains of the monsoon will come.

So will flooding. But 90 percent of the year's rainfall in this season means crops will again grow and feed the nation.

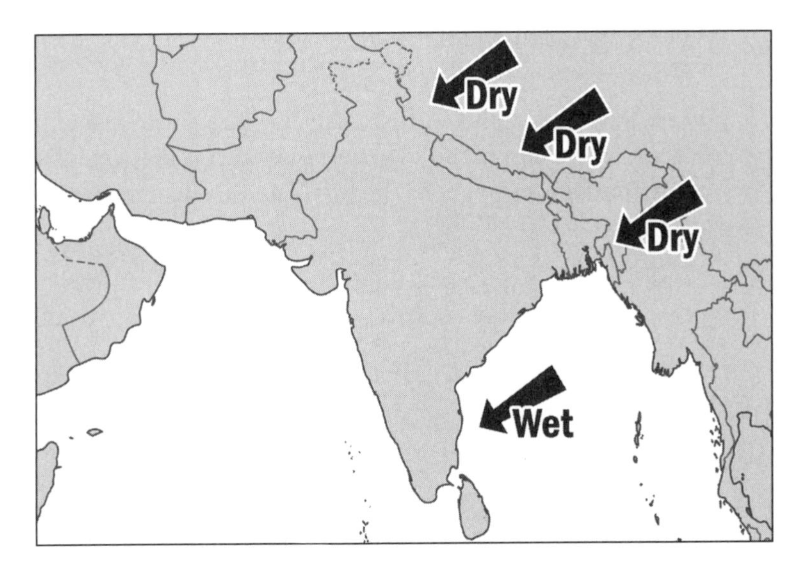


*Dry monsoon.*

*(Courtesy of the University of Nebraska-Lincoln School of Natural Resources.)*

## The Dry Season

The winter monsoon is much different. In fact, it's 180 degrees different. The winds now have shifted and blow from the northeast from November to April. If the wet monsoons don't provide enough water, people will starve during the dry monsoon.

The floods of the wet monsoon are nothing compared to the famine and disease the dry monsoon will see. In 1770, a year of drought caused a famine that killed an estimated 10 million people in Bengal. In 1866, one million people died in Orissa.

India has built irrigation systems and facilities to store water, reducing famine outbreaks.

Still, in 1998 a heat wave killed 2,500 people, most in the eastern state of Orissa. In fact, the dry monsoons can see some record hot temperatures along with drought.

More than 330 people lost their lives in June, 2005, when 137-degree temperatures scorched an already over-dry India.

Only the wet monsoon could stop the damage, which it did later that month.

Mahrashtra and southern Andhra Pradesh witnessed the deaths of more than 1,400 people in the dry monsoon of 1996.

**Storm Stats**

An estimated 10 million people died from famine because of the lack of rain during the wet monsoon in Bengal in 1770.

In 2002, more than 1,200 people died during the Indian dry monsoon. The toll was highest in Andhra Pradesh with 1,037 lives lost from the May temperatures that eclipsed 110 degrees.

Between 1899 and 1901, about 15 percent of the population in Gujarat, in the western portion of India, starved to death.

We think of the wet monsoons as the killers—their flooding can be catastrophic—but the summer dry monsoons can be apocalyptic in scope when the heat and dryness bear down.

## Wicked Weather

While the wet monsoon gives life with its nourishing rain, it also takes life away. Every year, roughly 1,000 people lose their lives to the monsoon's weather. With just over one billion people living in India—many in overpopulated cities—when the floods come, they claim many victims.

But while India has suffered tremendous human losses throughout history, it was the monsoon in Thailand from September to December 1983 that claimed the most lives: 10,000.

The number is staggering. The floodwaters also caused $400 million in damages, and 100,000 more people contracted waterborne diseases. To top it off, 15,000 people were forced from their homes.

India, of course, has seen its share of death and destruction. The historical record counts great casualty numbers.

The worst monsoon-induced flooding in 100 years, however, was in July 2005, when the death toll from monsoon rains topped 1,000.

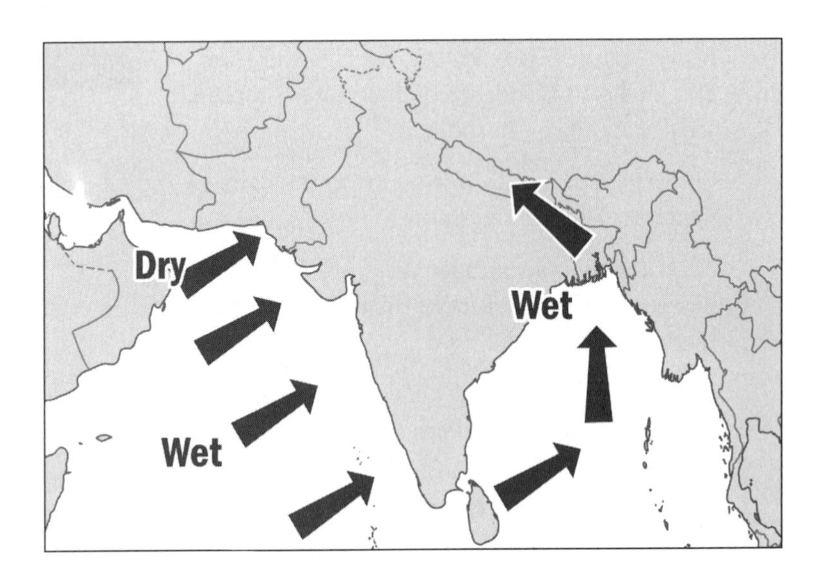


*Wet monsoon.*

*(Courtesy of the University of Nebraska-Lincoln School of Natural Resources.)*

**Storm Stats**

On average, 1,000 people are killed each year due to monsoon-related weather.

Mumbai, formerly Bombay, in eastern India was deluged by relentless rain that flooded the city and caused massive mudslides.

Rescue workers couldn't even approach the mudslide areas as the rain continued to fall in the financial capital of India. In fact, in just half a day, the city recorded more than half of its yearly rainfall.

Monsoon rains in Bangladesh were also particularly bad in 2004. Some 230 rivers run through the poor nation, and when flooding comes, it comes with a fury.

An estimated 600 people were killed by the July and August monsoon rains as mudslides and flooding, as well as disease in the water, took their toll. Millions of people were left homeless, and an estimated 25 million people had been affected by the flooding—the worst since 1998.

Damages from the July–August 2004 flooding totaled $6.6 billion, but the flooding that brought death and destruction to Bangladesh brought much of the same to more than 50 million people in surrounding nations.

Nearly 30 million people in India were affected by monsoon rains and subsequent flooding, and 482 people lost their lives in northeastern India alone.

In Nepal, more than 200 people died and nearly a million more were affected by the severe rains, flooding, and mudslides.

Monsoons may bring nourishment, but, unfortunately, there is a high price to pay.

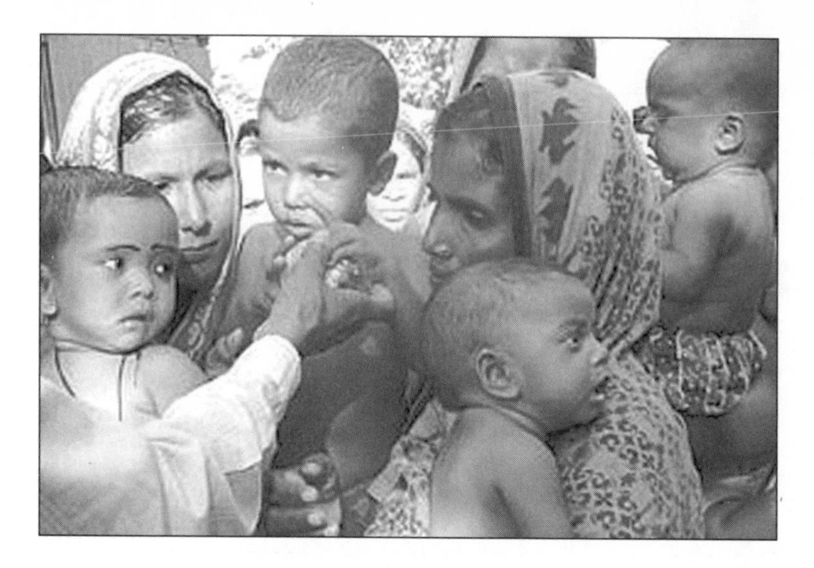

*Food is distributed to flood victims in Bangladesh. Millions of people would starve were it not for international relief aid during extreme storms.*

*(Courtesy of the United Nations World Food Program.)*

# The American Monsoon

Here's a phrase you never hear: "Hundreds of people have lost their lives in the worst monsoon in Arizona in decades."

Sounds odd, doesn't it? Most of the statement is comical—most of it. Monsoons really do happen in Arizona. They just don't typically bring death and destruction; remember, a monsoon is defined as any wind that reverses itself seasonally.

In North America, each year a smaller-scale monsoon—but a monsoon nevertheless—whips through Mexico and on up into New Mexico and Arizona.

During much of the summer, winds that once blew west to northeast over much of the southwest change to a more southeasterly direction, bringing with them wet air from the Pacific, Gulf of Mexico, and Gulf of California. Thunderstorms develop when that warm, moist air meets the cooler mountain air that rises. It's not uncommon for winds to gust to near 60 miles per hour in the more severe thunderstorms.

**Storm Stats**

Acapulco receives 55.1 inches of rain each year. But 51.8 inches of that occurs from June through October, the monsoon season.

**Storm Stats**

The average start of the Arizona monsoon is July 7, and it typically ends September 13.

Rain totals in Arizona, usually near the mountains, average 11.46 inches during the monsoon period. However, some severe monsoon storms have caused major damage in Arizona.

Maricopa County saw 115-mile-per-hour gusts on August 14, 1996. Damages in the county reached $160 million.

**Storm Stats**

The greatest number of monsoon days in Phoenix was 99 in 1984; the least was in 1962, with 27.

The Labor Day Storm of 1974 combined monsoon rains with heavy remnants from the Pacific Tropical Storm Norma. In its wake, massive flash flooding claimed the lives of 23 people.

Uncommon, but Arizona monsoons can pack a punch.

# A Rite of Passage

The monsoons that occur halfway around the world can be quite misunderstood. As we mentioned earlier in this chapter, monsoons are a way of life—for humans, animals, and all things that grow—and maintain a delicate balance not just in the areas most affected by the monsoon seasons, but all across the globe, as heat and cold are distributed through these powerful ebbing and flowing winds.

**Storm Stats**

More than 2 billion people from Africa to Asia depend on the wet monsoons for their water supply.

What the weather books and TV shows don't show is the culture, the legend, and the romance of the monsoon. Western cultures often are more engaged in the catastrophic—the droughts, floods, and mudslides.

But there is a beauty that is embraced, even celebrated, when the monsoon arrives.

The dancing peacock, with its beautiful plumage, has long been a symbol of the monsoon in folklore and culture. Festivals are held in anticipation of the wet season that will bring nourishment and life to the people. While the monsoons are seen in different lights in northern India versus the southern portion of the country, the oncoming season is one of celebration for all.

In fact, tourists even flock to cities like Kerala and Goa during the wet monsoon for "herbal rain holidays."

Indian art and song depicts, more than almost any other subject, the romance of the monsoons. Themes are common of people coupling up romantically with the music of the rain tapping on the roofs. It's a time of courtship, from royalty to the common man hustling on the streets.

**Inside the Storm**

The peacock is the symbol of fertility and good luck and is significant in the Indian monsoons.

Paintings and photographs capture the magic of lightning, and the replenishing effects of the rains. In literature, the rains wash away the dry and spring abundant life, from plants and grasses, to dragonflies and birds—almost overnight. The life-giving rain washes away the grime and dry dirt staining the cities and streets. It is a time to reflect and give thanks for life.

Not everyone sees the rain as totally life-sustaining. There is a great deal of fear and anticipation that comes with the seasonal change: Will it be torrential this year? Will it put me at risk? Will it come at all?

Like many phenomena worldwide, that which cannot be easily understood in these cycles of life becomes the stuff of prayer, hope, and sacrifice.

And no more does it show than in monsoon cultures.

## The Least You Need to Know

- There are two monsoons: a wet monsoon and a dry monsoon.
- Wicked monsoon rains can cause massive deaths from drowning and mudslides, but a season of drought can kill millions.
- North America sees a yearly monsoon, from Mexico up to northern Arizona.
- Much folklore and art embraces the monsoons in many cultures.

Chapter 16

# Destination: Rain Forest

## In This Chapter

- They don't call it a rain forest for nothing
- Getting acclimated
- The rainy city
- Lessons learned

The last frontier on Planet Earth is home to some of the most pronounced weather extremes. In terms of precipitation there is no place that consistently receives more rain than rain forests.

Plants, animals, and even people have somehow adapted to living under a perennial rain cloud and thrive. Fact is, while the world's rain forests cover only about 6 percent of the earth's land surface, they are home to more than half of the plants and animals on this planet.

In this chapter, we'll take a look at just how much rain falls in these appropriately named regions, how rain forest ecosystems are balanced, and what hangs in that balance as their resources are threatened and depleted.

# Rainiest Places on Earth

It rains a lot in many forests of South America. It rains in the deep woods of Asia, China, Washington State, and even Alaska. But does that make these regions rain forests?

Well, that depends. You see, there are two types of rain forests in this world. *Tropical* rain forests probably come to mind when we think of the big, wet jungles of South America. And you're dead-on. It rains all year in a tropical rain forest and it's in a tropical climate. But there are *temperate* rain forests, too, and Alaska's got one.

The difference in Alaska, though, is that the forest sees a season of rain rather than a whole year of rain. The precipitation, however, doesn't just end when the "rainy season" is over. There is quite a bit of fog, and that wet haze helps maintain the rain forest climate.

**def•i•ni•tion**

There are two types of rain forests: **tropical,** where it rains all year, and **temperate,** where it rains in a season but sees fog year-round.

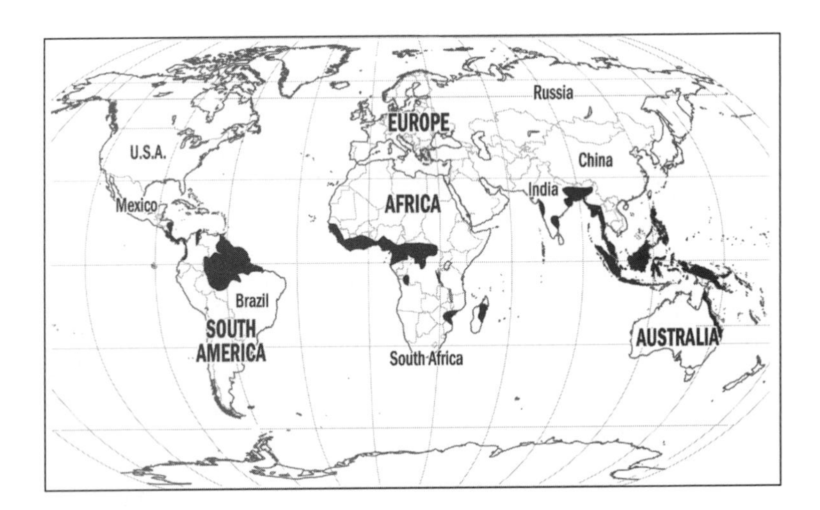


*Rain forests of the world.*

*(Courtesy of the Blue Planet Biomes.)*

We'll take a look at a few of those types of rain forests later in this chapter, but let's first look deep into the Amazon and just how extremely rainy it is.

## Heavy Rain

The Amazon is the granddaddy of all rain forests, covering the entire basin of the Amazon River (the second-longest river in the world). The rain forest receives about

9 feet of rain in an average year. To put that in context, the United States Eastern Seaboard states receive about 5 feet of rain a year, while the Plains states see a bit over 1 foot on average.

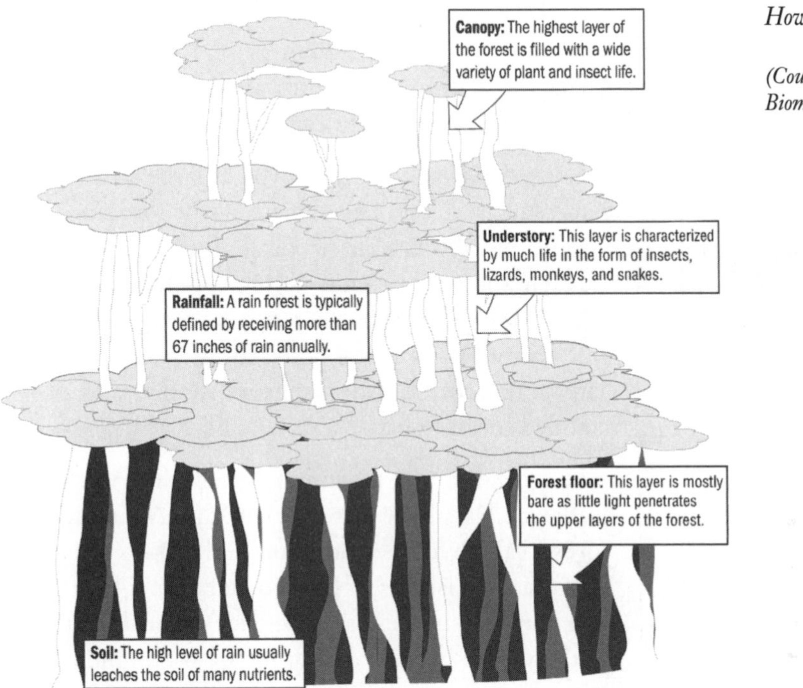


*How a rain forest works.*

*(Courtesy of the Blue Planet Biomes.)*

Because of its equatorial climate, it rains more—almost every day. The tropical ocean climate and its proximity, mixing cool air from all that evaporation with that warm seawater, produces great amounts of rain.

The high humidity occurs all year because of that rainfall. In fact, the tropical rain forest receives about an eighth of an inch of rain each day.

Here's exactly how it works: the hot equatorial sun warms the land and sea, and water from both evaporates into the air. As that warm, moist air rises it cools at higher elevations and condenses to form droplets. The droplets combine, get too heavy to be in the vaporous clouds, and shabam! Rain.

**Eye of the Storm**

The Amazon is the largest rain forest in the world and is home to 30 percent of all animal and plant life on the planet.

The average temperature in most rain forests is 77 degrees Fahrenheit. The temperature never drops below 64 degrees and seldom crests above 93 degrees.

## Where the Wild Things Are

Tropical rain forests also can be found in Central America, in Zaire, Africa, as well as some western African areas and Madagascar, in Indo-Malaysia and in New Guinea, and Queensland, Australia. Almost all of these places are equatorial.

Much of the rain forests in Central America have been destroyed as large areas have been clear-cut for cattle ranching and sugar caning. The climate is ripe for rain forests, but this region's tropical rain forests have become sparse. The same is true of Madagascar in Africa. Most of this rain forest has been destroyed, but central Africa is still home to the world's second-largest rain forest, with mangrove swamps and flooded forest floors.

**Storm Stats**

World rain forests produce about 40 percent of the earth's oxygen.

**Eye of the Storm**

Rain forests once covered 14 percent of the earth's land surface. Now, they cover 6 percent. Experts estimate the last remaining rain forests could disappear before 2050.

The second-largest mangrove forest in the world is in Bangladesh, believe it or not. Other Asian rain forests can be found from India and Burma to Malaysia and islands of Borneo and Java. The heat and humidity in these areas keep it tropical for most of the year, especially during the monsoon months, but a drier period does prevail during the winter monsoon.

Australia also owns a chunk of the world's tropical rain forests on the Pacific side of the continent. New Zealand and New Guinea, once combined with each other and Australia millions of years ago, also share rain forests.

Southeast Asia's 3,100-mile chain of 20,000 islands between Australia and Asia also is home to some pretty dense rain forests.

The wet monsoon seasons help control the temperature and humidity and it even rains each day during the dry monsoon season. Temperatures stay relatively constant all year, at about 80 degrees Fahrenheit. Some of the main countries in this island chain include the Philippines, Singapore, Thailand, Vietnam, Cambodia, Indonesia, Laos, and Brunei. Each year, these areas receive 100 inches or more of rain.

**Storm Stats**

The Amazon River basin was formed between 200 million and 500 million years ago.

# American Rain Forests

If you thought that only tropical rain forests have extreme rains, think again. The temperate rain forests of North America see a whole lot of rain every year—enough to classify them as rain forests.

And that's saying something, especially when we think of the stereotypical hot, balmy, mosquito- and crocodile-infested jungles of South America and Borneo. And that's a stark contrast from snowy, mountainous Alaska.

But the Northwest Pacific rim, from Washington State on up to lower Alaska, holds quite a treasure of ecology: the temperate rain forest.

## Washington State

Unlike the broad, leafy trees and plants of the equatorial rain forests, the conifer rules in the Pacific Northwest. Washington's rain forest is packed with conifers. Whatever broadleaf trees exist in a temperate rain forest are deciduous, meaning they lose their leaves each year and regrow new leaves in the springtime.

But just like the tropical rain forests, these temperate areas receive gobs of rain each year—only seasonally. For much of the year it will rain in Washington's rain forest. And when it's not raining, like in the warmer summer months, it's foggy. It's this fog that allows moisture to be present in the forest, enabling the flora to flourish.

**Storm Stats**

The Olympic Peninsula is home to some natural marvels. Eight types of plants and 15 animals are found nowhere else on earth.

Here's how it happens: moisture-rich air from the Pacific rises, just like it does anywhere else, but in Washington, that air becomes trapped by the coastal mountains and condenses. When that happens, the rain falls. And it does so often. If you've ever visited Olympic National Park in the northwestern portion of Washington State, you'll have experienced this firsthand.

**Inside the Storm**

A temperate rain forest must be in a mild coastal climate, have heavy summer fog, and see a very generous rainfall.

*Great pines stand blanketed in rich moisture in Olympia National Park in Washington State.*

*(Courtesy of the National Parks Service.)*

For the most part, a temperate rain forest will see an average of up to 60 inches of rain each year. Olympia sees more like 140 inches. Temperatures in this rain forest can drop below freezing, but it's uncommon. In the summertime the temps hardly ever top 80 degrees.

**def•i•ni•tion**

**Old-growth** forests have developed undisturbed over long periods and are often referred to as virgin forests.

All that rain and fog helps the trees grow tall, wide, and old. In fact, Washington State is home to some of the oldest *old-growth* stands in the country.

The temperate rain forest of Washington State isn't unlike that which is found in Alaska's Tongass National Forest.

## Alaska

Alaska may be the last place you'd think there would be a rain forest, but the cold, snowy state is home to an incredible temperate rain forest inside Tongass National Forest, which was established in 1907 by President Theodore Roosevelt.

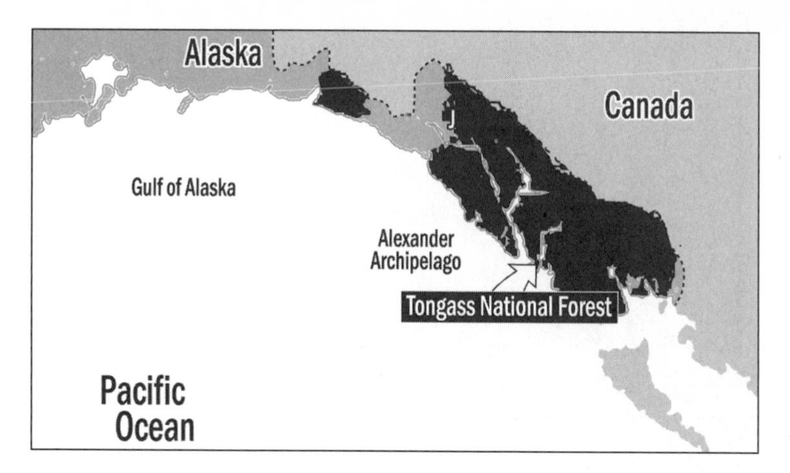


*Tongass National Forest.*

*(Courtesy of the National Oceanic and Atmospheric Administration [NOAA].)*

What's so cool about Tongass? Well, for one it's in Alaska, and if that's not good enough, it is the largest intact temperate rain forest in the whole wide world. Its land mass consumes an amazing 17 million acres in part of the region called the Alexander Archipelago. The area stretches some 500 miles north to south and includes more than 100,000 islands.

The trees—giant sitka spruce—were growing for hundreds of years before Roosevelt protected them. Rainfall in Tongass often reaches 100 inches a year.

People still live in Tongass—some 75,000 Alaskans, many of them descendants of tribes who've lived there for centuries.

But not only humans call Tongass home. More than 3,000 eagles flock to the rain forest to fish and breed. Year-round, nearly 400 bald eagles live there.

Although Tongass is protected, mining and logging have ripped apart millions of acres of temperate rain forests worldwide. The southern half of North America's temperate rain forest used to be the world's biggest. Now, as much as 95 percent of it has been logged. Alaska now accounts for more than 40 percent of the world's remaining temperate rain forests. About 12 percent of the Alaskan rain forest has been clear-cut.

*Home to some five unique salmon species, the Tongass National Forest comprises 40 percent of the world's remaining temperate rain forests.*

*(Courtesy of the National Oceanic and Atmospheric Administration [NOAA].)*

## How Nature Adapts to Extreme Rain

No matter where you go around the world, people, animals, and plants somehow adapt to even the harshest of conditions.

One week in a rain forest, and you'll have called it quits.

But some living things can be quite resilient—even in spite of a deluge almost every day.

It would seem that the human populations inside rain forests would be sparse. Not so. Some 50 million tribal people live in rain forests worldwide.

### People Really Live Here?

The Yanomami, which means human being, is the largest group of Amerindian people in South America.

Living among the fertile rain forests of southern Venezuela and northern Brazil, this tribe sustains itself on hunting, fishing, and farming, to a lesser degree. Like most tribes, their survival greatly depends on keeping their environment intact.

For generations, these warrior-hunters have sustained themselves amid harsh conditions, growing vegetables, like sweet potatoes, in the moist, nutrient-rich soil.

Even today, some 20,000 Yanomami live in that culture, although it's no longer as primitive in many senses.

But as mining and logging encroach on their land and culture, their spiritual beliefs become imperiled. Some even kill themselves over it. Others have been slain as those mining and logging operations move in.

Largely untouched before 1980, by 1987, about 2,000 of the Yanomami had been killed over their land.

> **Storm Stats**
>
> There were an estimated 10 million Indians living in the Amazon rain forest five centuries ago. Today there are fewer than 200,000.

The pygmies of central Africa are another tribe that for generations have survived untouched by outsiders. Hunter-gatherers, these tribes have lived in harmony with the rain forest without damaging its delicate ecosystem.

A neat fact about the pygmies, besides that they are some of the shortest people on earth at about 4 feet tall, is that they require very little water. Most of the food they eat is packed with water. They also sweat much less, since the moisture-rich air cools their bodies without having to sweat.

This group packs up camp whenever the resources in their areas get scarce. The dense jungle offers a bounty of hunting and gathering opportunities.

Like the Yanomami, the rain forest that they live in is being destroyed by mining and logging.

The governments have tried to teach the pygmies to farm outside of the rain forest, but they've been resistant.

## Animal Kingdom

Like humans, animals have adapted to the wet environment of the rain forest. Birds, reptiles, amphibians, and even monkeys have found the rain forest habitat, well, hospitable.

> **Storm Stats**
>
> At more than 1 billion acres, if the Amazon rain forest were a country it would rank ninth in size.

Even New World monkeys have used evolution to get a leg—err, a tail up. The species have adapted to tree life by using prehensile tails—tails that can literally grasp tree branches as they swing from tree to tree.

All the animals have adapted their eating patterns to consume heavy amounts of fruits—and insects.

**Storm Stats**

One in five birds in the world lives in the Amazon rain forest.

Insects comprise the largest single group of animals in the tropical rain forests.

Experts estimate that the Amazon rain forest is losing 137 plant, animal, and insect species every day because of deforestation. That staggering number equals 50,000 species a year.

## Rooting for Plants

There is no other place on earth that contains a wider variety of plant and animal life than the Amazon rain forest. Ranking a distant number two are the rain forests of Southeast Asia. In any tropical rain forest, there can be as many as 100 different species to every 2.5 acres.

**Storm Stats**

About 25 percent of all the medicines we use come from rain forest plants. Lymphocytic leukemia patients have a 99 percent chance of remission because of the rain forest's rosy periwinkle plant.

Rain forest plants also have had to adapt through time to the harsh, wet conditions that would kill lesser species.

To shed water more quickly, the leaves have formed waxy coatings, grooves, and so-called drip tips, preventing the weight from snapping the stem or branch.

The plants still need broad leaves, though, to soak up as much sun as possible, since not much of it ever hits the rain forest floor.

Those plants have proved to be great in their medicinal value, too. Worldwide, 121 prescription drugs sold come from plants. About a quarter of Western pharmaceuticals contain rain forest ingredients. However, only about 1 percent of the tropical trees and plants in the rain forests have been tested by scientists.

While there is a lot of hope that cures for cancer and AIDS might reside in these untested plants, more and more scientists and ecologists worry that those plants and trees will be gone before they can be used in labs. The real killer, they predict, will be more deforestation.

One last note, however, is that those trees and plants in the Amazon rain forest actually help your quality of life—in the way of clean air. More than 20 percent of the world's oxygen is produced in the Amazon rain forest.

**Eye of the Storm**

The Venus flytrap is one of the many plants that has adapted to rain forest life. With the plentiful insect population, this plant has taken to eating bugs—and small reptiles!

## Dying Rain Forests

We hear it all the time: "The rain forests are dying."

Funny how so many people have heard it so often that they've become desensitized to it. The facts have been there in black and white for some time. "Yeah, but it's so big."

It is. So big, so dense, so remote .... Unfortunately, those three factors could be what is allowing for so much of its destruction. Policing the rain forests in the Amazon alone would take the work of several countries—many of them still developing, corrupt, or without any sort of military or way to prevent rain forest destruction.

Mining operations, timber industries, and global warming are all major contributors to the deforestation of the rain forests.

Despite the cries for responsible practices and government intervention, the destruction continues and, in some cases, has even increased.

**Storm Stats**

In the last 15 years, nearly 100,000 square miles of the Amazon rain forest have been deforested.

According to satellite images from the National Institute for Space Research, the speed of deforestation in the Amazon rain forest has increased from 2001 to 2002 by 40 percent. That figure was the highest rate of deforestation since 1995. The institute figures show that there were 7,010 square miles lost to deforestation in 2001, and in 2002, another 9,840 square miles were lost. Illegal farming, mining, and timbering were to blame.

Simply put, destroying the rain forest means not only destroying the plants, animals, and people who live there, but destroying chances to test more medicines, better understand our planet, and keep our air and water quality pure.

Global warming already is taking its toll on the rain forest canopies, but without the vast cooling effect that the rain forests provide, the planet will only get warmer.

> **Storm Stats**
>
> Forested acreage the size of Poland is lost from the rain forest every year.

A warmer, thus stormier, planet means more severe monsoons, hurricanes, cold weather, heat, and drought. It also means soil erosion, crop production, fishing, hydroelectric generation all suffer as well. Carbon dioxide levels not being filtered through the trees' and plants' leaves can bring a whole other level of worry.

The situation is happening in the Amazon rain forest at the same time other nations are struggling to keep their natural resources intact—and not just in rain forests.

Deforestation in Central America was to blame for the extreme damage brought on by Tropical Storm Mitch in 1998. When the storm parked over El Salvador, Nicaragua, and Honduras, the heavy rains that once could have been in part absorbed, instead caused massive flooding and mudslides. When Mitch left the region, more than 10,000 people were dead.

## The Least You Need to Know

- There are two types of rain forests: temperate and tropical.
- Alaska has the largest temperate rain forest on the planet.
- Half of the world's plant and animal population lives in rain forests, which comprise just 6 percent of the planet.
- Deforestation is killing the rain forests, which could be completely destroyed by 2050.

Chapter 17

# En"lightning" Encounters

## In This Chapter

- Ingredients for lightning
- Big bang theory
- Lightning strikes twice
- Strike survival

The Empire State Building is struck by lightning an average of 25 times a year. The queen of the New York City skyline was hit so often she was used for years as a lightning observatory. The building offers a direct challenge to that old wives' tale; lightning, indeed, can strike twice.

Objects conducive to electricity—homes, cows, people—can be hit by a bolt more than a couple of times.

## Lightning Extremes

Every year, about 100-Americans on average are killed and hundreds more are injured in lightning strikes.

No small wonder: lightning packs a 100-million-volt wallop, reaches temperatures of about 50,000 degrees, and travels 5 miles in seconds.

**Inside the Storm**

A bolt of lightning can contain as much as 1 billion volts and 200,000 amps—enough to light a 100-watt lightbulb for three months!

Chances are, if you're living on Planet Earth, at any given moment there are approximately 1,800 thunderstorms raging. Add to that statistic that the average number of lightning strikes around the world every second is 100, and the picture becomes pretty clear: lightning is all around us.

And it's a killer. For every 86,000 flashes in the United States, one person dies from a lightning strike.

*Multiple cloud-to-ground and cloud-to-cloud lightning strikes during a nighttime thunderstorm.*

*(Courtesy of the National Oceanic and Atmospheric Administration [NOAA] Photo Library.)*

Sounds like doomsday, right? It doesn't have to be. We'll let you in on a few tips on how to avoid being struck later in this chapter, but for now, take comfort in this: the odds of a person being struck in the United States are about 300,000 to one. The odds of being killed by lightning in the United States are closer to three million to one.

But even with those odds, lightning remains the number-two weather-related killer of Americans. Only floods kill more people.

## How Lightning Forms

Believe it: the common ingredient to making lightning is probably the last thing you'd think of. It's ice. Without ice in the clouds, there would be no Thor, no Zeus, no Shazam! And forget about the T'bird.

When a thunderstorm forms, it does so with moisture, instability, and cold. When a cold front and a warm front meet, great instability results 6 to 10 miles above the earth's surface. Clouds form, and the cold air in them mixes with the newly formed moisture in the higher parts of the clouds, forming ice. When all that ice—tiny positively charged crystals and negatively charged hail—gets to whipping around up there, rising and sinking, the product is electricity.

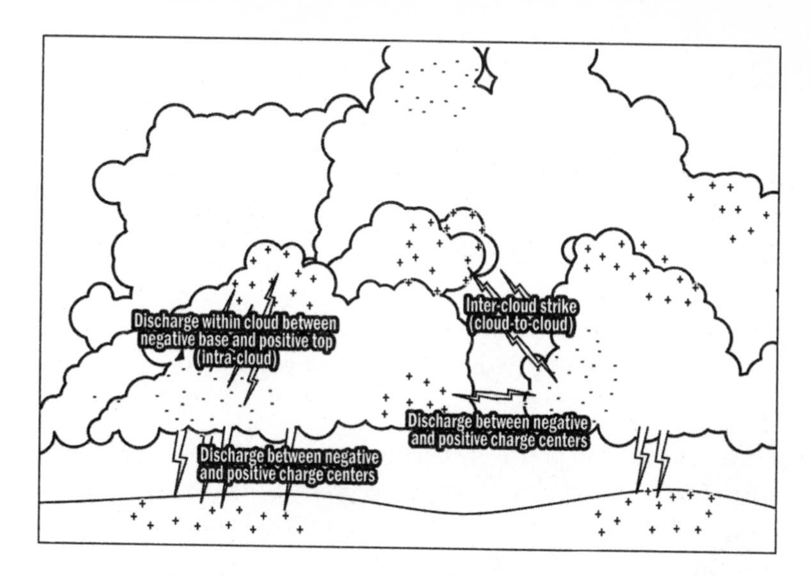


*Lightning at night.*

*(Courtesy of National Weather Service.)*

All that energy needs to go somewhere, and there are a few different paths it can take. We'll look at the different forms of lightning in a moment, but for now, know that the energy is seeking places to go. Tall points on the ground—church steeples, telephone poles, the Empire State Building—are great nearby targets for that charge.

When the negatively charged *stepped leader* is attracted to the positively charged particles on the ground, a conduit of electricity forms—and this is what we see as lightning.

The bolts of lightning have been known to strike more than 50 miles away from a storm. It might be a blue sky over your home, then wham! Lightning. In the air, lightning can travel more than 100 miles! Most commonly, though, the bolt takes the nearer path to an object, usually in the summer, when great instability between hot planet and cold front is more prevalent.

**def•i•ni•tion**

A **stepped leader** is a negatively charged electrical "feeler" that moves in steps looking for a positively charged object to use as a conduit.

When lightning strikes an object, it can travel a great distance on the ground—as much as 60 feet. Plumbing, pipes, phone lines, and vent stacks all are good conduits for lightning to travel through. Your house may get struck, and your buddy's TV next door receives a surge during the football game. It happens.

# Types of Lightning

Lightning is a pretty fickle beast. Its travels and formation can be trivial, whether it travels cloud-to-cloud, cloud-to-ground and back up, ground-to-cloud. Scientists have spent years deciphering its code. Cloud-to-cloud is the most common type of lightning, but there are some pretty cool variations up in the clouds that myths and legends were built upon.

Cloud-to-ground lightning is the most destructive, but there are two sides to this story. There's a negative flash and a positive flash, the latter being even more destructive.

The negative flash happens between the positively charged ground and the negatively charged clouds at the lower part of the storm. Positive flashes occur between the upper-level positively charged part of the storm and the negatively charged ground or area surrounding the storm. The negative strike travels cloud-to-ground, and the positive ground-to-cloud. The connection between the two generates the light we see—what we believe to be lightning. It happens in seconds.

In the positive flash, the channel between the two opposite charges forms in the anvil of the storm and comes straight down. The negatively charged stepped leader comes from the ground upward to meet the positive charge in the clouds. When the two meet as the positive reaches down, the flash happens. Boom!

**Storm Stats**

The approximate economic damage that lightning inflicts in the United States alone each year is a whopping $5 billion.

## Forked Lightning

This type of lightning is usually what comes to mind when we say the word *lightning*. These are the bolts shooting forklike out of ominous clouds to the ground, but they also can travel cloud-to-cloud—the most common path. About 20 percent of all of these strikes hit the ground.

## Sheet Lightning

This form of lightning happens within a cloud and looks like the entire cloud is lit up when it forms. There's nothing unique about this form of lightning—it's pretty ordinary.

## Heat Lightning

It rhymes with "sheet lightning" if you're ever writing a poem, but that's about all that these forms of lightning share. This lightning form is lightning from a storm a long way away—more than 10 miles away, to be precise. It's seen most commonly on hot summer nights when there's a clear sky above. Somewhere fairly nearby, someone is heading for cover.

*A nighttime thunderstorm brings typical forked lightning to Norman, Oklahoma, in March 1978.*

*(Courtesy of the National Oceanic and Atmospheric Administration [NOAA] Photo Library.)*

## Ball Lightning

This type of lightning is not very common at all. In fact, it's so odd that scientists dub it a phenomenon. It looks like an illuminated orb of lightning, just floating through the sky fast or slow. It can make a hissing sound or can be silent. Sometimes it ends with a bang. But the things have only been witnessed, and never photographed. Folks on airplanes have reported them, as well as others on the ground who claim the fireballs have come clean through their windows, leaving a burn mark in their wake. Some scientists call it hogwash, but there are a good number of believers.

## St. Elmo's Fire

This form of lightning is real, but is pretty freaky. The positively charged electric field glows green-blue usually just above an object, like a church steeple, a sailboat mast, or the horns of a bull. The soft glow arcs continuously and heads into the sky to seek out a negatively charged source in the clouds. Sailors named it after their patron saint because it looks like some great force is actually pushing it away.

### High-Altitude Lightning

Far atop a thunderstorm, flashes of colorful light—red, green, blue—emanate when regular old lightning is discharged from a storm cloud. There are some funny shapes and names that go with them, based on their colors: green elves look like glowing blobs; blue jets are trumpet-horn shaped and streak out of the cloud tops; and red sprites zoom upward from the cloud, sort of pointy in shape.

## Extreme Thunderstorms

In 1991, a Canadian *severe thunderstorm* toppled an entire forest—500 square miles, completely blown down! Thunderstorms can get pretty extreme. We're all taught to head for cover when that thundercloud roars to life or fires its first warning shot of lightning, but there is plenty of peripheral damage from the wind, lightning, hail, and rain that a thunderstorm can bring.

**def•i•ni•tion**

A **severe thunderstorm** is a thunderstorm that contains one or more instances of hail greater than ¾ inches, winds stronger than 57 miles per hour, or a tornado.

That was the case in Melbourne, Australia, on December 3, 2003. Two thunderstorms parked themselves over the metropolitan area with severe wind, flash flooding, frequent lightning, and hailstones the size of softballs! The rain caused massive flooding, winds toppled trees and buildings, and the hailstones fell for an amazing 20 minutes, causing damage to an estimated hundreds of vehicles. More than 39 inches of rain fell in the greater Melbourne area.

The storm was the worst to hit Australia's second-largest city in nearly 30 years, causing an estimated $100 million in damages.

Not all thunderstorms are these wicked tempests, but they can become very powerful. In fact, most of the time they are fairly small. But that doesn't mean they should be taken lightly. After all, lightning kills.

So we've all seen what a thunderstorm can do, but let's break it down. How does that bone-jarring clap of noise happen, and why?

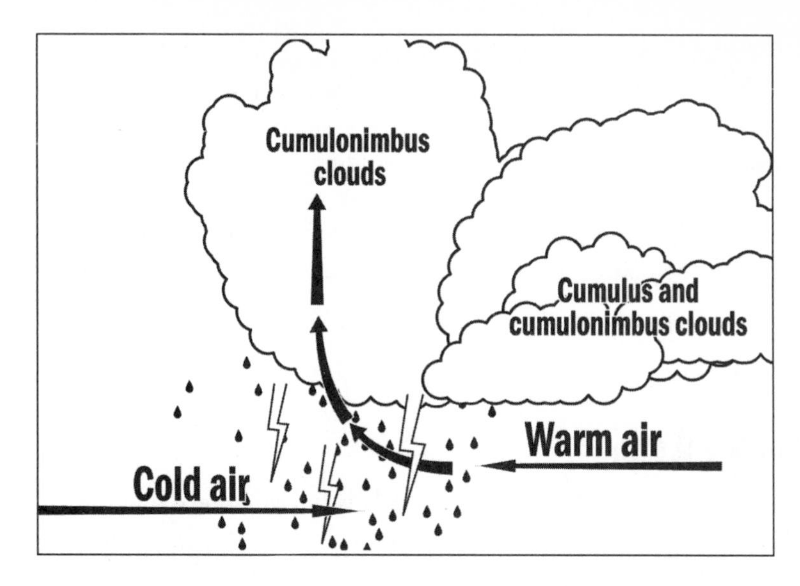


*How lightning forms.*

*(Courtesy of the National Oceanic and Atmospheric Administration [NOAA].)*

It's pretty simple. When lightning strikes, it moves in a path that's a balmy 50,000 degrees. The extreme temperature from that bolt forces a fast expansion of heated air and BA-BOOM! Thunder. Remember that lightning is traveling way faster than the speed of sound, so if you hear the thunder as you see the lightning, well, you might not live to tell about it—it will be right on top of you.

So how does that storm form again? The answer is about the same as most other storms: warm air rises upon evaporation, meets cold air aloft, usually in a front, the evaporated particles gather and form clouds or they fall and the pressure gradient between the warm and cold air causes wind. The cold air aloft forms ice particles out of the moisture, and those crystals whip around and become charged. Within a few minutes you get lightning, then thunder.

**Storm Stats**

Taking cover beneath trees during a thunderstorm is a bad idea. In fact, it's the third leading cause of death during a lightning storm.

# Hair-Raising Events

Being hit by lightning, as we've seen, can happen just about anywhere and at any time to anyone. With so many lightning strikes around the planet during every season, it's a wonder we all don't know someone who's been zapped.

But it's not all a game of chance. Sure, you could be barbecuing chicken legs at the state park barefoot under a perfectly blue sky and the next thing you know, you're lying flat on your back from a lightning strike, but that's pretty rare. Folks who get hit by lightning are usually out in it—on a golf course, holding on to a golf club that acts as a lightning rod, or taking cover under a big, leafy oak tree .... So the odds of you being killed by a bolt of lightning if you're an average Joe are about 3 million to one; the odds of the average Joe being struck and not killed are about 300,000 to one.

But if you live, say, in the so-called lightning capital of the world, you're no longer so average.

## The Lightning Capital of the World

Now, back to that 5-iron acting as a lightning rod. Florida is full of golfers, and the golf season never ends. Ditto with just about any outdoor activity in this warm-all-year-round state. Tennis, boating, sunbathing, soccer .... If you're living in the Sunshine State, chances are you're retired and are outside a heck of a lot more than folks in Dubuque, Iowa. And you're no longer so average.

Add to that Florida's composition. The peninsula is surrounded by warm water, has lots of moisture in the air, is low and flat—basically a thunderstorm magnet.

Its geography, composition, and population make it the Lightning Capital of the World.

The area stretching from Orlando to Tampa, a less than two-hour drive on a good day, is where most lightning strikes occur.

**Eye of the Storm**

Folks in Tampa know they live in the Lightning Capital. In fact, they even capitalized on the odd trait when they named their National Hockey League team the Tampa Bay Lightning.

So if you're contemplating the move to Florida after retirement, consider this: the odds of being struck by lightning are now 100,000 to one, and the chances of being killed by a thunderbolt is closer to one million to one.

Remember, now, the population of Florida is about 17 million. Do the math ....

For the most part, the mainland of the United States sees fewer lightning strikes than Florida, but there are even fewer heading toward the Northwest. Florida's atmosphere carries a lot of moisture below 5,000 feet, and very warm ocean temperatures generate afternoon sea breezes that move that moisture-rich air around.

**Lightning deaths by state from 1995-2004**

*(Courtesy of Storm Data.)*

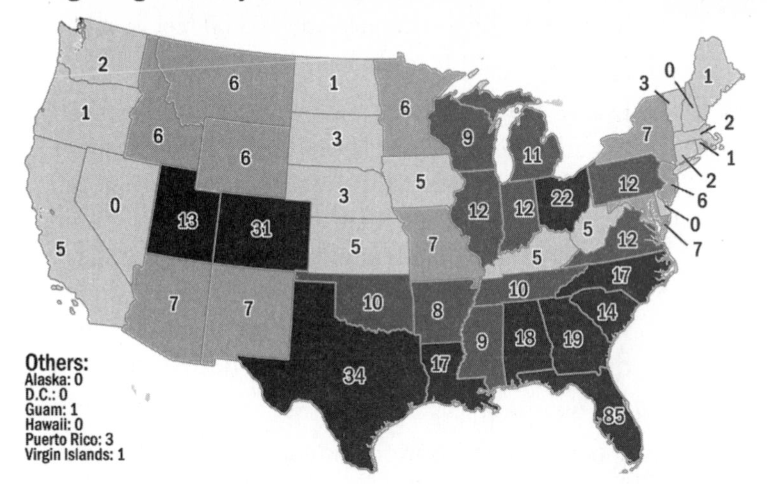


Truth is, from 1990 to 2003 Florida saw 136 lightning deaths, ranking it number one in that category. In contrast, Alaska had no lightning fatalities. But no other state even comes close to matching Florida. Texas ranks second on the list with 52 deaths, followed by Colorado with 39. The total for the United States in this period is 756 deaths related to lightning.

**Eye of the Storm**

A Navy ammunition depot in New Jersey was hit by lightning in 1926. The depot exploded, killing 19 people and injuring 38 others.

Men are more likely to be struck than are women. The stats for the period from 1959 to 1994 show that 84 percent of those hit were men, while 16 percent were women. Most of those strikes came in the summer months of June, July, and August. Again, outdoor activity plays a large part in your chances of being struck.

# I've Been Hit!

Obviously, you want to avoid being hit by lightning. No good can come from being struck by a thunderbolt. There are some pretty obvious ways to avoid becoming a lightning statistic, but it's probably fair to say that those who have been struck by lightning weren't out looking for it. They are, after all, accidents.

And in some instances it can't be avoided: being stuck in a severe thunderstorm on the rush-hour commute through Orlando, caught in a sports stadium in Dallas,

> **Storm Stats**
>
> Lightning wreaked havoc on more than 220 British tall ships during the Napoleonic wars, but the British Navy refused to install lightning rods, since they were invented by the rebel Benjamin Franklin.

hunkering down inside your Nashville, Tennessee, home, traveling in an airplane 10,000 feet over Atlanta. Really, it could happen anywhere.

In fact, lightning has caused a whole bunch of fatalities and injuries in airplanes: 68 were killed when a TWA flight over Milan, Italy, was hit during its climb; 16 died when their plane was hit by lightning over France; 73 died on initial approach to Philadelphia (this one exploded the fuel tank); 86 died when their plane was hit over Peru, catching the right wing on fire and plunging it into the mountains.

The National Oceanic and Atmospheric Administration has compiled a survey of people who were struck by lightning but lived to tell their tales. They have a support group named Lightning Strike and Electric Shock Survivors, International.

One person was struck when talking on a cordless telephone inside her house. She was temporarily paralyzed in her lower body. A man was out fishing when he was struck. The lightning caused burns over 60 percent of his body. Another man was completely paralyzed when lightning hit him on a Cape Cod golf course. Another man was working on his car in his garage when lightning hit him. He has burns, pain, and short-term memory loss.

We cannot always control the situation, but with a good dose of common sense, we certainly can alter it.

## Avoiding a Strike

Shelter is key in an electrical storm. The best shelters are those that are grounded, that is, have some sort of plumbing, pipes, or electrical wiring that will absorb and transfer a lightning hit. Your house is a great example of a good shelter, while your tool shed, assuming it's a typical free-standing structure without plumbing and wiring, is a bad example. The latter is still better than no shelter at all, but if you have the choice, go with the house.

Cars offer pretty good protection from a lightning strike, too. It's the metal and wiring that are key, not the rubber tires. Rubber has nothing to do with the equation. And cars make better shelters than ungrounded tool sheds, but still not as good as your house.

*Lightning strikes the ground near Norman, Oklahoma, during a 1980 thunderstorm.*

*(Courtesy of the National Oceanic and Atmospheric Administration [NOAA] Photo Library.)*

The 30/30 rule is the one to tuck away in your brain. It's a very good defensive plan against being struck by lightning. So if you see lightning, count on the next bolt being close by. Using the 30/30 flash-to-bang rule, you count the seconds between the flash of lightning and the bang of thunder. If it's within 30 seconds, the storm is too close and you need to take cover. It will be safe to go outside again only 30 minutes after the last 30-second flash to boom.

Wait a half hour before venturing back out.

Avoidance is the best way to not get hit by lightning. Don't be out in it. Before you head out on that kayak trip, take a look at the weather forecast. Chance of a thunderstorm? Don't paddle off 20 miles through the marsh. And if you are outside, take a look around you. Are there dark clouds on the horizon? Can you feel or see an increasing wind? Do you hear thunder? If you can say yes to any one of those questions, guess what? You are within range of being struck by lightning. Take cover. It can be a deadly mistake to say, "Let's wait 'til it gets closer."

**Eye of the Storm**

Thirteen people were injured by lightning at an outdoor rock concert in Baltimore, Maryland, in 1998, despite the installed lightning rods at the venue.

Shelter, as we've said, can be the safety of your home, a nearby building, or even your automobile. Stay away from tall objects, like trees, utility poles, and church steeples. If you're inside, unplug appliances and stay off the phone.

Showering is also a pretty bad idea, since lightning has been known to travel throughout a home's plumbing. (And being soaking wet wouldn't be a good thing either.) Stay away from the plumbing altogether, if you can help it. It's a good excuse to put off doing the morning dishes, anyhow.

Air-conditioning units should be turned off, too.

If you can't make it to a safe shelter, and you're stuck outside when a storm hits, remember: no trees. If you're in a boat or swimming in the lake or ocean, you need to get ashore and find shelter.

Find a low-lying open place away from trees and poles and anything made out of metal, then crouch down low to the ground—hands on knees, head tucked between them. Whatever you do, do not lie down flat on the ground. This makes you a bigger target, keeping in mind that the lightning shoots back up from the ground, as we've learned, and the greater the surface area, the greater the target you are.

**Storm Stats**

More Americans are killed by lightning than any other type of storm except floods.

If you are with a group, spread out to make yourselves less of a target.

If someone is struck by lightning, get medical help immediately. People who have been struck are safe to handle. Be careful when you handle them, though, as they may be badly burned where they were struck and where the electricity exited their body. Provide rescue breathing if their breathing has stopped. If their heart has stopped, a trained person should administer CPR.

## The Least You Need to Know

- Every year, about 100 Americans on average are killed and hundreds more are injured in lightning strikes.
- Some forms of lightning are more dangerous than others, but all should be avoided.
- Lightning is the most underrated weather killer. Taking shelter in a storm is paramount.

# Part 5

# The Heat Is On

They call it the silent killer. Heat and humidity's one-two punch can sneak up on people with little warning and knock them out of the game. Record heat waves and dry seasons have claimed countless lives and ruined whole tracts of lands via desertification.

In this part of the book, we'll visit some dreadfully hot places, look at severe heat waves, and outline how folks should keep their cool.

# Chapter 18

# Hotter Than the Blazes: Extreme Heat Waves

## In This Chapter

- What's hot, what's not
- Deserts and other hot places
- Summer sizzler
- Feel the burn

Death Valley. The name alone conjures up an image of old, intact bones of some beast stopped dead in its tracks by the relentless burning of the sun. A lone vulture circles overhead ….

You'd be right to invoke that image. Death Valley is home to the hottest temperature on record in North America. In July 1963, the desert reached a balmy 134 degrees Fahrenheit.

Still, it was second to El Azizia, Libya, where it reached 136 degrees in 1922.

It's not only deserts that get hot. Cities feel the effects, too, and they even see heat waves that sizzle in summer months.

And the heat takes lives and sometimes costs billions.

## When You're Hot, You're Hot

We'll check back in on El Azizia and Death Valley in a few moments, but first things first: what's hot and how does it get that way?

What's obvious to us is that it's hottest in the summertime, when the sun is closest to us in the Northern Hemisphere. The days are longer, too, allowing more hours of sun and thermal heating.

What's not so obvious is how the weather patterns shape up for a hot summer.

### Heat's Highs and Lows

In summer, the Bermuda High sets up in the Atlantic Ocean. As it moves into the central Atlantic, it spreads quiet weather through a good portion of the central and eastern United States. With that Bermuda High in place, storms usually don't form in those parts of the country. Instead, we get stagnant weather patterns. The high will set up and stay in the area for extended periods of time, allowing temperatures to heat up under dry conditions. These are the types of systems that can give us the so-called *dog days* of summer where hot days.

**def•i•ni•tion**

**Dog days** are named after Sirius, the Dog Star, and its close position to Earth from July 3 to August 11.

The Pacific Northwest also can see a high-pressure system set up in that region in summer that allows for similar hot, dry patterns to exist. As high pressure builds in an area for a few days, the ground begins to dry out, even if rain has recently fallen. As the ground gets drier, it allows temperatures to climb higher. Drier ground can heat up faster than moist or saturated ground. So you can see that the temperatures on the third day of a heat pattern will be a few degrees higher than on the first or second day.

**Storm Stats**

The longest hot spell on record was in Marble Bar, Australia, where it was 100 degrees Fahrenheit or higher for 162 consecutive days, from October 20, 1923, to April 7, 1924.

Summertime is full of high-pressure systems. They bring warm, dry air to a region and tend to park there for days, sometimes weeks.

During the daytime, the sun's rays heat the ground and the ground absorbs heat. At night, if skies are clear and winds are light, any heat that was absorbed into the ground is radiated back into space. There's nothing to hold the heat in the ground. This is called radiational cooling. However, if there are clouds present at night, the clouds act as an insulator to keep the heat close to the ground.

Also, if the winds rise at night, they will help prevent the temperature from dropping off. The winds keep the relatively warmer air from the day stirred up and don't allow it to radiate back to space. Once you lose the winds and the cloud cover, the heat goes back out into space.

It can still feel warm on a cloudy day; however, you can notice the difference in the type of warmth. A warm, humid, cloudy day feels more uncomfortable than a hot sunny day. It's still hot, but it's a drier heat versus a more humid heat. When you sweat on a drier day, the perspiration evaporates off your skin, cooling you off. On a more humid day your body is not as readily able to evaporate the perspiration it makes, so you remain a little hotter.

Also, you can still get sunburn on a cloudy day because ultraviolet radiation penetrates clouds. So even though it doesn't feel as hot or look as sunny on a cloudy day compared to a sunny day, you are still getting the UV rays that can cause sunburn.

**Eye of the Storm**

The *Guinness Book of World Records* lists Dallol, Ethiopia, with the highest average annual mean temperature at 94 degrees Fahrenheit.

Meteorologists define heat spells by looking at surface and upper-level charts that would show an area of high pressure entering a region and not moving much, especially in summer. That way they would know that a period of dry, quiet, and possibly hot weather is on the way.

## Down in the Valley

Weather conditions in valleys can last longer than similar weather conditions over flat surfaces. Cold temperatures and fog can get trapped in what's called "cold air damming," where cold air enters a mountainous region and gets dammed up in the valleys because it has nowhere else to go. It slides down into the valleys and doesn't have any wind to stir it up or move it out.

The same can be said for hot temperatures. The heat can get trapped in the valleys. It will feel a bit unbearable because the winds die off in valleys. That wind would typically provide a breeze on a flat surface and cool you down a bit. In valleys, there really isn't good airflow because the mountains cut it off.

An area can also stay overheated because of a lingering warm front, a condition in which temperatures rise gradually over a long period of time. When a warm front enters an area, the leading edge of the warmer air overrides the colder air at the surface. The cold air retreats to the north as the warmer air comes in from the south.

*A farm field bakes under the hot summer sun during the Dust Bowl era in 1938.*

*(Courtesy of the National Weather Service.)*

The warm air riding up and over the cold air creates clouds at the frontal boundary and well ahead of the surface warm front. This creates a more gradual slope with the warm front. The first signs of an approaching warm front are cirrus clouds. They develop well in advance of the warm front location. Any rainfall associated with warm fronts is usually lighter and covers a greater area for a longer period of time. As the warm front passes, there is a gradual rise in temperature. Warm fronts also move slower—about 15 miles per hour.

The earth's tilt, warm fronts, and high-pressure systems show us how it gets hot and stays hot, but how about extremely hot? The cook-an-egg-on-the-sidewalk kind of hot? The kind of hot that stops a large beast in its tracks?

# Just Deserts

Good ol' Planet Earth is about one third land, the rest water. But a third of that third of land is desert.

Believe it or not, there are several types of desert—cool coastal deserts, cold winter deserts, and even polar deserts, like in the Antarctic. But when we think of deserts, we think of the sand, dunes, bones, and that vulture circling overhead. Oh, yeah—and extreme heat.

You can bank on the 12 *subtropical deserts* of the world being hot, dry, and deadly. And, if you were to bet on it, you'd be correct to say that the hottest temperature ever recorded was in a desert.

**def•i•ni•tion**

A **subtropical desert** is one that receives fewer than 10 inches of rain a year.

Interestingly, most of the 12 deserts are jammed between 5 and 30 degrees north and south of the equator. That's where the name *subtropical* comes from.

## How Deserts Form

Climate has everything to do with why we have deserts. It's hot year-round in the subtropical belt where deserts exist. And you can bet it's dry, too. And when you mix hot and dry, nothing grows. When nothing grows, you have no evaporation, no moisture raining down, and the cycle continues.

We're closest to the sun around the equator—in fact, the sun is at a 90-degree angle here, which creates extreme high-pressure zones. The heat is literally piled up, so that it's hot above and below the equator. It's just hot all around. The hot air rises, but there's also hot air on the ground and just above it. No sea breeze, no cold front, no evaporation.

That allows the sun to just bake, bake, bake all day. And, as we've learned, with no cold front, no evaporation, and no clouds, the heat escapes into a starry, starry night, every night. Deserts can get a bit nippy after midnight.

Thermometers measure the air's temperature, but can you imagine how hot the surface gets? Park your minivan in the Gobi Desert and the temperature on the auto's hood would definitely exceed 150 degrees—frying an egg would be no problem. Heck, you might even fry bacon, too (not recommended).

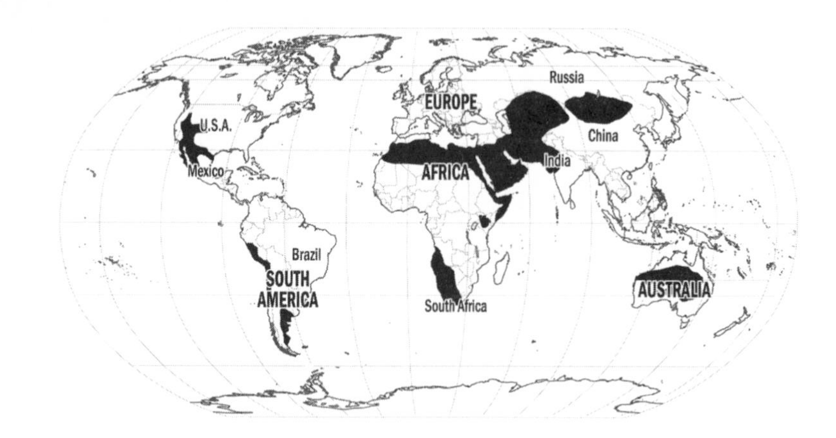


*Deserts of the world.*

*(Courtesy of Extremescience.com.)*

## Desert Extremes

So let's get back to El Azizia, Libya, where it was 136 degrees in 1922, and Death Valley, where it was 134 in 1963, and visit some other extremely hot places.

> **Storm Stats**
>
> Most of the 12 subtropical deserts of the world are within 5 degrees and 30 degrees north and south of the equator.

It's important to note here that when we talk about extreme heat, as with extreme cold, there is some debate about who had the hottest temperatures. The difference between El Azizia's 136 degrees and Death Valley's 134 degrees is pretty small, and it's safe to say that there were probably hotter places on earth at some point. There aren't thermometers stationed in every square mile in the Sahara, for instance.

And there are a lot of deserts. For all intents and purposes, though, we'll call Death Valley and Libya the record holders, since most of the weather information in the world tends to gravitate to those two spots.

Just survey some of the extreme temperatures in the world, and most of the hottest on record are in or near the deserts.

### Highest Temperature Extremes

| Rank | Temp (F) | Place | Date |
|---|---|---|---|
| 1 | 136 | El Azizia, Libya | 13 Sep 1922 |
| 2 | 134 | Death Valley, California | 10 Jul 1913 |
| 3 | 129 | Tirat Tsvi, Israel | 22 Jun 1942 |

| Rank | Temp (F) | Place | Date |
|---|---|---|---|
| 4 | 128 | Cloncurry, Australia | 16 Jan 1889 |
| 5 | 122 | Seville, Spain | 04 Aug 1881 |
| 6 | 120 | Rivadavia, Agentina | 11 Dec 1905 |
| 7 | 108 | Tuguergari, Philippines | 29 Apr 1912 |

Hot stuff, to say the least.

Most of Death Valley resides in the eastern-central portion of California. It's name is appropriate. Let's tackle the "valley" part first: it's the lowest elevation in the Western Hemisphere at 282 feet below sea level. The "death" part comes from a heroic attempt in 1849 by 30 men trying to find a shortcut through it to get to California. Of those 30, 18 made it. The rest? Bones, vultures … you get the picture.

But Death Valley shares a lot in common with Libya. Add Iraq and Algeria, too. These are the hottest places on earth.

**Storm Stats**

Between July 6 and August 17, 1917, the temperature in Death Valley was above 120 degrees every day.

# Heat Wave Happenings

Roughly 1,000 Americans lose their lives to heat-related illness each year, and all those unfortunate numbers aren't exactly happening in the Mojave Desert.

Lots of times those lost lives occur in the cities—the infirm, the elderly, and the very young. Air movement is hampered in these concrete jungles that absorb and store heat, block wind with massive man-made structures, and trap the heat with smog, not allowing it back into the atmosphere, day or night.

Solar radiation killed 20,000 people in the United States in the 40-year period of 1936 to 1975.

Those numbers have declined with the advent of more affordable air-conditioning units and mandatory "brown-outs," but with global warming and greater populations that put more demands on electricity, those numbers may rise again.

**Eye of the Storm**

More than 14,000 people died in France during a heat wave that plagued all of Europe in 2003. An estimated 50,000 people were killed throughout Europe.

But what causes a heat wave? High pressure stuck in park. These high-pressure systems combine their hot temperatures with humidity, and their duration and lack of movement make life, well, unbearable.

So let's take a look at a few disastrous heat waves.

The eight days between August 31 and September 7, 1955, in Los Angeles killed 946 people. The temperature through that stretch averaged 100 degrees.

But that's nothing. An estimated 10,000 people died in 1980 when a heat wave parked over the central and eastern United States between June and September. The heat wave caused about $20 billion in damages.

Just eight years later, as many as 10,000 people died in a heat wave that struck the same areas of the United States. Damages totaled $40 billion.

Around the world, heat waves are just as lethal. In 2003, one of the worst heat waves ever struck Europe. Total deaths were estimated at 50,000, with France being hit hard with 14,847 deaths.

One of the reasons for so many fatalities was that it doesn't usually get that hot in France. People not familiar with 100-degree heat were unprepared. Lack of air-conditioning was a major factor, especially with the infirm and elderly populations. Most of the deaths occurred in this segment of people.

Italy also saw a great number of deaths—an estimated 20,000. For weeks, temperatures stayed between 100 and 104 degrees Fahrenheit. In the United Kingdom, 100-degree sustained temperatures killed 907 people. At Heathrow Airport in London on August 1, the temperature hit 100 degrees, the highest temperature there since records began in 1911.

The heat wave gripped the entire European region, and those in cities fared the worst. However, the peripheral effects of the heat wave claimed many lives as well. For instance, 18 people died in Portugal when an estimated 10 percent of that country's forests were consumed in fire attributed to the heat wave. And the damages were not measured only in loss of human life: Germany's Danube and Elbe rivers were unnavigable by ships, and the shipping trade took a major hit. Drought was prevalent throughout the continent. Flash floods from glacier melts in the Swiss Alps caused chaos in Switzerland when temperatures soared to 106 degrees. A staggering 75 percent of the wheat crop production in the Ukraine and 80 percent in Moldova were stunted.

**Storm Stats**

The United States has had some severe heat waves. During the Chicago heat wave of July 14 through 20, 1995, 739 people died.

# Beat the Heat

Heat kills more people than any other weather extreme in the world. The human body tries to adjust to heat—you can feel it; you can't always see it.

When a tornado is bearing down on you, you run for shelter. When the cold weather causes your breath to crystallize right in front of your nose, you head indoors. But for some reason, we think we can beat the heat.

*The sun sets behind a windmill, ending a hot day near Houston, Texas, in May 1972.*

*(Courtesy of the Environmental Protection Agency [EPA].)*

Not so. People need to take precautions and be smart about heat, especially if a heat wave parks itself over a city.

**Eye of the Storm**

The best way to beat the heat is to stay out of it.

The best way to beat the heat is to—duh—not be in it. Stay indoors with good ventilation—preferably with air-conditioning—and drink plenty of liquids.

Check in on the senior folks in your family or neighborhood. Those who are sick, too. These people are exceptionally susceptible to the heat and might be at risk.

You should avoid drinking alcohol and caffeine in heat as they tend to dehydrate you.

Remember, it's the sustained heat, not just the few minutes that you're out in it, that kills. If you don't have air conditioning, spend the hot afternoon at the public museum, where it's artificially cooled.

If you're outside, wear lightweight, loose-fitting, light-colored clothing and cover up exposed skin. Always wear sunblock.

If you must work outside, drink a lot of water, take frequent breaks, and be sure you're not alone.

And for heaven's sake, save mowing the lawn for a cooler day. Chances are, it won't grow in this heat anyway!

## The Least You Need to Know

- Hot air can park over an area with little warning and cause great discomfort, and even death.
- The European heat wave of 2003 was among the worst ever.
- Thousands of people die each year due to extreme heat.
- The best way to beat the heat is to not be in it.

Chapter 19

# "Real Feel": Humidity's Role

## In This Chapter

- What causes extreme humidity
- Close call
- Where the humidity hides
- Turning up the heat

We call it muggy weather. That oppressive, in-your-face dense air that seems to cover everything like a hot, wet blanket.

And let's not even talk about what it does to naturally curly hair ….

But humidity is more than just a blanket of hot, wet air. It can make life uncomfortable—even unbearable, causing health problems and death when mixed with the heat of the day. In this chapter, we'll look at humidity's role on the planet, some of the most humid places on earth, and how we can avoid getting "mugged."

## Sticky Situation

We couldn't live without humidity. We might want a little less of it on those dog days in July and August, but that humidity is what keeps the planet working. What you're witnessing on a hot, humid day is the very life cycle—and in many cases, you can literally see it at work.

Normally, humidity can't be seen—it's a water vapor, and that means zillions of particles being heated and turning into gas and rising into the atmosphere. Just don't tell that to folks in Baltimore, Maryland, who "see" some of the most humid air each hot summer day.

## Cause and Effect

First of all, humidity is basically a measure of *water vapor* in the air. The higher the number, the more humid it is.

**def•i•ni•tion**

**Water vapor** is water in a gas phase. It can be produced by the evaporation of liquid water or the sublimation of ice.

In a nutshell, the basics of evaporation, as we've talked about throughout this book, apply. Water falls to earth, is stored in living things—plants, trees, and even you and me—and in bodies of water (liquid, like the ocean, or solid, like a glacier).

When the sun beats down, that water evaporates—that is, turns into a gaseous state—where it rejoins the other vapor in the heavens, groups together, becomes too heavy to remain in its cloud state, and falls back to earth. Repeat the process an infinite number of times and that's the vapor cycle.

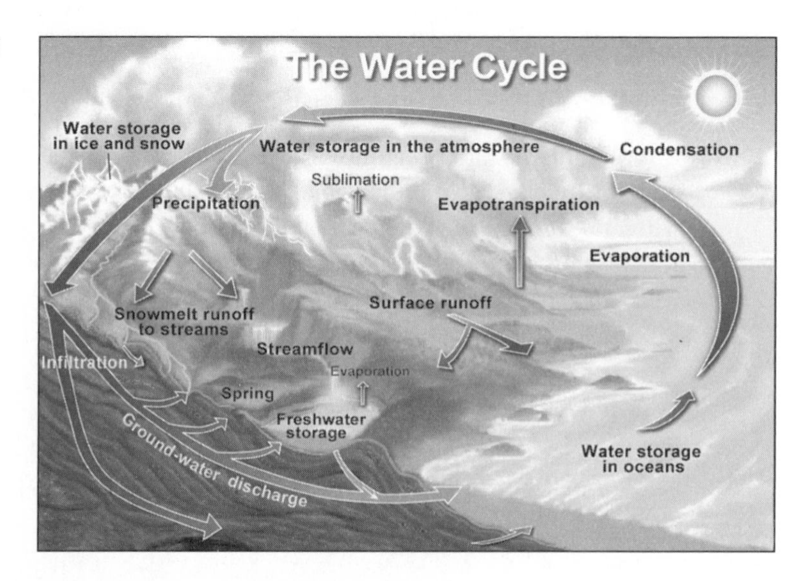


*(Courtesy of the U.S. Geological Survey.)*

But there are a few other factors, such as deserts. Not really humid there. That's because there is very little water on the ground to evaporate. A random cactus won't cut it. Wind also plays a part. When it's a dry wind blowing off a desert in a high-pressure system, it's going to be pretty dry.

It's a rule of thumb.

If you live in a rain forest, it's going to be one of the most humid places you'll ever experience. Ditto with the southeastern United States. So many swamps and much closer to the equator than New York City.

And ol' Baltimore? Just think of all that water surrounding a fairly warm city. It's a lot more humid near the Baltimore Harbor than Omaha, Nebraska.

**Storm Stats**

Tropical rain forests are among the most humid places on earth, with average humidity levels at 77 to 88 percent.

Saturated air can't take on any more water, and it begins to form quite a strong, thick layer of common vapor.

Eventually, thank goodness, it reaches a saturation point and has to release all that water. Clouds form, rain comes.

But then the whole cycle repeats if it's still hot out, and it usually is all of August, especially in Baltimore.

## Absolutely, Specifically Relative Measurements

There are actually three different ways to measure humidity in this world. There are relative, specific, and absolute humidity.

Relative humidity is the one we commonly hear—it's a percentage. It's the ratio of the current vapor pressure of water in a gas (like air) to the vapor pressure when the gas becomes saturated.

Absolute humidity is the mass of water vapor in the air or gas, by weight, usually measured in grams per cubic meter or cubic foot.

**def•i•ni•tion**

The **dew point** of the air is the temperature that the parcel of air must be cooled to at a constant barometric pressure in order for the water vapor to condense into water, a.k.a "dew."

Last, specific humidity measures the amount of aqueous vapor in the air comparing that vapor to dry air.

The last part of this equation that you're probably semi-familiar with is the *dew point*. We see dew on the ground on most warm mornings.

# How Does It Feel?

There are a whole lot of ways to measure humidity. Specialized scientific equipment collects data and measures pressure and water, vapor and air. But just by going outside and taking a deep breath, you can tell if it's humid out.

> **Inside the Storm**
>
> The first hygrometer to measure humidity was made with human hair in 1783 by Horace Benedict Saussure. The oil in the hair made it curl or straighten with the humidity. Modern hygrometers use semiconductor materials to measure humidity.

Still, a cool trick is the so-called bad hair day method. Really. For eons scientists have relied on this method. No, they didn't use a long-haired townswoman to watch the effects; but they did use horse hair strands to watch whether they curled or straightened, and measured it.

So, when you're having a bad hair day, it's because humidity actually curls one's hair. Conversely, when it's not humid, your naturally curly hair will be, well, naturally curly.

Sweat from the skin is another good indicator of high humidity. But what's really happening is moisture in the air is acting as your sweat and your body may not produce sweat, which can be dangerous, since it's not really cooling off as much as it would if you were sweating. If that really gets severe, the blood that normally comes to the surface to cause the sweat also stops flowing to the brain and muscles. In severe cases, the condition is called hyperthermia, or heatstroke. We talk a lot more about that in Chapter 20.

*Heat index.*

*(Courtesy of the National Weather Service.)*

**Relative percent humidity**

| Temperature | 40 | 45 | 50 | 55 | 60 | 65 | 70 | 75 | 80 | 85 | 90 | 95 | 100 |
|---|---|---|---|---|---|---|---|---|---|---|---|---|---|
| 110°F | 136 | ←Dry and hot | | | | | | | | | | | |
| 108 | 130 | 137 | | | | | | | | | | | |
| 106 | 124 | 130 | 137 | | | | | | | | | | |
| 104 | 119 | 124 | 131 | 137 | | | | | | | | | |
| 102 | 114 | 119 | 124 | 130 | 137 | | | | | | | | |
| 100 | 109 | 114 | 118 | 124 | 129 | 136 | | | | | | | |
| 98 | 105 | 109 | 113 | 117 | 123 | 128 | 134 | | | | | | |
| 96 | 101 | 104 | 108 | 112 | 116 | 121 | 126 | 132 | | | | | |
| 94 | 97 | 100 | 103 | 106 | 110 | 114 | 119 | 124 | 129 | 135 | | | Humid and hot ↓ |
| 92 | 94 | 96 | 99 | 101 | 105 | 108 | 112 | 116 | 121 | 126 | 131 | | |
| 90 | 91 | 93 | 95 | 97 | 100 | 103 | 106 | 109 | 113 | 117 | 122 | 127 | 132 |
| 88 | 88 | 89 | 91 | 93 | 95 | 98 | 100 | 103 | 106 | 110 | 113 | 117 | 121 |
| 86 | 85 | 87 | 88 | 89 | 91 | 93 | 95 | 97 | 100 | 102 | 105 | 108 | 112 |
| 84 | 83 | 84 | 85 | 86 | 88 | 89 | 90 | 92 | 94 | 96 | 98 | 100 | 103 |
| 82 | 81 | 82 | 83 | 84 | 84 | 85 | 86 | 88 | 89 | 90 | 91 | 93 | 95 |
| 80 | 80 | 80 | 81 | 81 | 82 | 82 | 83 | 84 | 84 | 85 | 86 | 86 | 87 |

So what's a comfortable level of humidity for humans and pets? A relative humidity of 30 percent to 60 percent seems to do very well for us. Since we feel hotter at higher humidity, then a drier air will feel more comfortable. If that humidity gets too low, though, such as in air-conditioned buildings with dehumidifiers, folks tend to feel cold to the bone, so some humidity is a good thing.

Outdoors, that humidity and temperature combo is measured by the heat index. Take a look at the previous graphic and you can see what's ideal and what's not.

**Inside the Storm**

There are 326 million cubic miles of water on Earth with only 3,100 cubic miles of it in the air, as clouds, vapor, or precipitation.

When that heat and humidity team up, it can be devastating. In 2003, Europe was struck with record-breaking heat and humidity, causing the deaths of 50,000 people.

We'll visit some humid places next, but remember, when the humidity climbs on hot days, take cover.

Even for folks in good shape, the heat-humidity combo can kill.

There's a false sense of perspiration going on, as we've mentioned above, and knowing the signs and being proactive can keep you healthy. We will discuss heat cramps, heat exhaustion, and heatstroke in the next chapter.

# Humid Places

Time for a quiz: Where is the most humid place on earth?

It's a trick question. The answer is the farthest from what most people think of as humid. Ready?

The Arctic.

It's so humid in the Arctic that it's actually one of the world's seven seas. It's so humid that it's wet—albeit frozen.

And it's bigger than the United States.

Rain forests have nothing on the Arctic. So how does that happen? Isn't it the evaporation from the sun and the heat that causes humidity?

**Eye of the Storm**

The Arctic, bigger than the entire United States, is the most humid place on earth.

Sort of. See, the Arctic waters are warmer than the air, and the ice is salty. And because that water is warmer, it creates a fog that rises up and covers the entire area in a balmy humidity that's second to none.

**Storm Stats**

The highest dew point ever recorded was in Dhahran, Saudi Arabia, at 95 degrees Fahrenheit on July 8, 2003.

But as far as regular "dry" land goes, Aseb, Eritrea, in the South Red Sea has an average dew point that often surpasses 84 degrees Fahrenheit.

*Latent heat* is released when the water vapor over the Red Sea evaporates, causing clouds and rain and very high humidity levels. The warm Red Sea waters produce tons of water vapor, increasing the humidity throughout the Red Sea Basin and usually manifesting itself as fog. All the humidity keeps the region, which includes the Eritrean and Yemeni highlands, very rich in moisture.

**def•i•ni•tion**

**Latent heat** is the amount of heat absorbed or released by a substance before changing states, such as water changing to steam, at a constant pressure and temperature.

Now, what about those rain forests?

Sure, rain forests are some of the most immensely intensely humid places on earth. The Amazon rain forest, as we learned from Chapter 16, receives about 9 feet of rain every year. Half of that rain returns to the atmosphere through evaporating and transpiring through plants and trees. Add to that rainwater the snowmelt from the Andes, with water levels rising as much as 45 feet, and millions of acres of rain forest are lush, water-packed lands just waiting for that warm sun and temperature to evaporate it. The result is a misty, humid environment.

# Hotter, More Humid

It seems that all we hear these days (and have for some time) is that *global warming* will be the end of us all. Fossil fuel–burning engines, among other factors, emit carbon dioxide—so-called greenhouse gases—which cause a dome of poisonous vapor in the atmosphere that traps more greenhouse gases and keeps the heat from exiting into the atmosphere.

Humidity plays a great role in that equation.

And that wet air traps the earth's heat like a nylon sock on a sweaty foot.

You see, water vapor is a greenhouse gas, just like carbon dioxide is.

**def•i•ni•tion**

**Global warming** occurs when the average temperature of the earth increases enough to cause a climate change.

So if you've ever been in a smoggy city on a humid day, you know health warnings are issued, and it's not a good idea to suck in a whole lot of that air; especially refrain from jogging or any strenuous activity.

Over a long period of time, though, all that trapping of gas is causing the planet to become warmer; melting glaciers, fewer cooling periods, and changes in air currents may be doing us in.

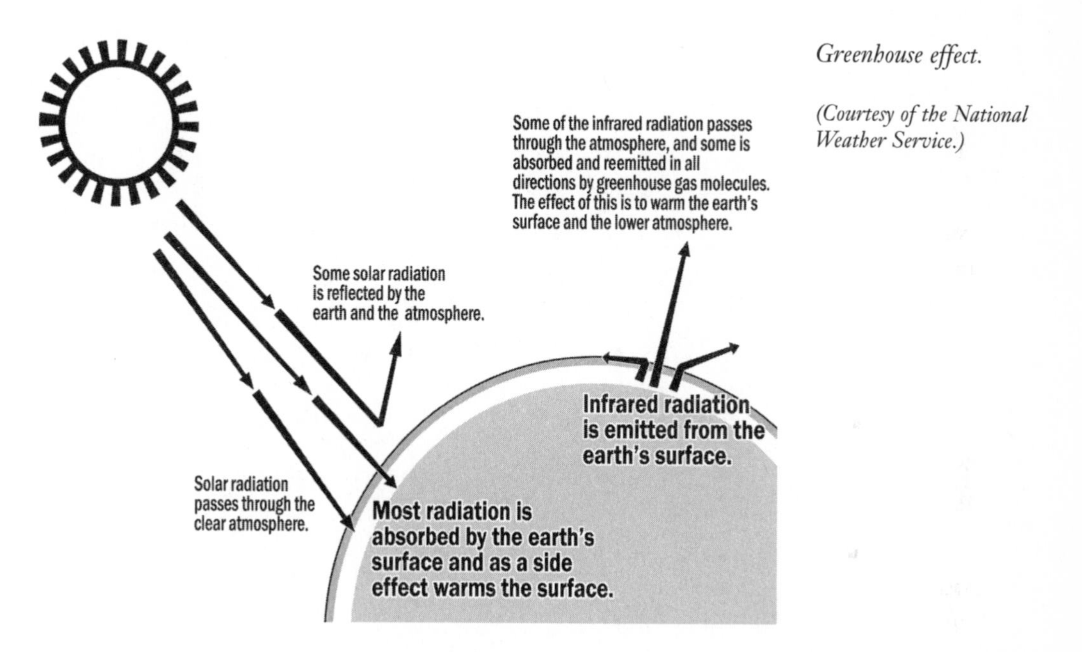


*Greenhouse effect.*

*(Courtesy of the National Weather Service.)*

So, the bad news with global warming is the more the planet warms, the more humidity we're going to see.

The whole process is called "positive water vapor feedback." There's a debate over whether a warmer planet causes more evaporation, but recent work by NASA scientists prove the theory is viable, although how much vapor is increased may not be as much as many had thought. But it's a start.

## The Least You Need to Know

- There are three ways to measure humidity: relative, absolute, and specific.
- Humidity and heat can cause detrimental effects on humans.

- The Arctic region is the most humid place on earth.
- Humidity's role in trapping greenhouse gases is stirring the global warming debate.

# Chapter 20

# Keeping Your Cool

## In This Chapter

- Too hot to trot
- Chill out
- Pets, plants, protection
- A world overheating

As we've learned in the last couple of chapters, heat and humidity can team up to throw a deadly one-two punch. Heat kills, and that's a fact. Like hypothermia and frostbite, heat-related illnesses can sneak up on a person. Mowing the lawn on a hot sunny day can produce light heat cramps or full-blown fatal heatstroke. Long-term illnesses can include some types of skin cancer.

Flora and fauna also are susceptible to heat. And the planet in general will be a much cooler place if we don't damage it.

In this chapter, we'll take a look at how hot weather extremes affect our otherwise temperate lives.

# Heat Kills

We look for blue sky. We look for sun. We are so conditioned to bellyache when it's rainy or cold that it seems the only time we enjoy the outdoors is when the sun is out and the temperatures are warm. Even the weather guy on TV tends to sigh when he forecasts more snow, or *rain* … you can hear it in his voice.

Well, stands to reason, doesn't it? Who wants to shiver in the cold rain?

So we tend to celebrate the sunny days by doing the 100,000 chores on that sunny Saturday, go for that 10-miler when it's "shorts weather," and hang out on that boat with another mai tai 'til sunset. In other words, we overdo it. And we get sick. Some of us even buy the farm.

In fact, more than 1,000 Americans die from heat each year.

## The Heat Is On

We know from the previous chapters why it gets hot. That pesky heat index tells the story.

Take another look at it.

*Heat index.*

*(Courtesy of the National Weather Service.)*

**Relative percent humidity**

| Temperature | 40 | 45 | 50 | 55 | 60 | 65 | 70 | 75 | 80 | 85 | 90 | 95 | 100 |
|---|---|---|---|---|---|---|---|---|---|---|---|---|---|
| 110°F | 136 | ← Dry and hot | | | | | | | | | | | |
| 108 | 130 | 137 | | | | | | | | | | | |
| 106 | 124 | 130 | 137 | | | | | | | | | | |
| 104 | 119 | 124 | 131 | 137 | | | | | | | | | |
| 102 | 114 | 119 | 124 | 130 | 137 | | | | | | | | |
| 100 | 109 | 114 | 118 | 124 | 129 | 136 | | | | | | | |
| 98 | 105 | 109 | 113 | 117 | 123 | 128 | 134 | | | | | | |
| 96 | 101 | 104 | 108 | 112 | 116 | 121 | 126 | 132 | | | | | |
| 94 | 97 | 100 | 103 | 106 | 110 | 114 | 119 | 124 | 129 | 135 | | | Humid and hot ↓ |
| 92 | 94 | 96 | 99 | 101 | 105 | 108 | 112 | 116 | 121 | 126 | 131 | | |
| 90 | 91 | 93 | 95 | 97 | 100 | 103 | 106 | 109 | 113 | 117 | 122 | 127 | 132 |
| 88 | 88 | 89 | 91 | 93 | 95 | 98 | 100 | 103 | 106 | 110 | 113 | 117 | 121 |
| 86 | 85 | 87 | 88 | 89 | 91 | 93 | 95 | 97 | 100 | 102 | 105 | 108 | 112 |
| 84 | 83 | 84 | 85 | 86 | 88 | 89 | 90 | 92 | 94 | 96 | 98 | 100 | 103 |
| 82 | 81 | 82 | 83 | 84 | 84 | 85 | 86 | 88 | 89 | 90 | 91 | 93 | 95 |
| 80 | 80 | 80 | 81 | 81 | 82 | 82 | 83 | 84 | 84 | 85 | 86 | 86 | 87 |

There you have it: high humidity plus high temperatures equals extremely hot weather. But it's what that weather does to you that causes problems.

A heat wave in 1980 killed 1,250 people. How does that happen? Heat kills because it can tax the human body—especially older people, babies, or the infirm—beyond its abilities. From 1936 to 1975, an estimated 20,000 people in the United States alone died from heat-related causes and *solar radiation*.

**def•i•ni•tion**

**Solar radiation** refers to the radiant energy of the sun, and comes in three rays: electromagnetic, infrared, and ultraviolet.

**80-90°F+: Caution**
Fatigue possible with prolonged exposure and/or physical activity.

**90-105°F+: Extreme caution**
Sunstroke, muscle cramps, and/or heat exhaustion possible with prolonged exposure and/or physical activity.

**105-129°F+: Danger**
Sunstroke, muscle cramps, and/or heat exhaustion likely. Heatstroke possible with prolonged exposure and/or physical activity.

**130°F+: Extreme danger**
Heatstroke or sunstroke likely.

*High risk.*

*(Courtesy of the National Weather Service.)*

The National Weather Service issues advisories for excessive heat. There are Excessive Heat Watches, Excessive Heat Advisories, Excessive Heat Warnings, Heat Warnings, and Heat Advisories.

An Excessive Heat Watch is issued 12 to 24 hours in advance of the onset of heat and to show the potential impact of the heat and to suggest precautions. Heed them.

An Excessive Heat Advisory is issued when the conditions produce a nondangerous situation that is a major inconvenience, and, like the watch, includes the areas affected, but also the potential temperatures, heat index, general impacts, and timing of events.

The Excessive Heat Warning is issued when conditions are considered dangerous. It includes the area affected, potential temperatures, heat index, general impacts and conditions, and timing of events, as well as precautions.

> **Storm Stats**
>
> The temperature inside a car in the summer can rise 3 degrees every 5 minutes and can reach temperatures of 150 degrees in an hour—even if the windows are open.

A Heat Advisory is issued when heat indexes develop within 24 hours of 100 to 105 degrees for one day or longer.

A Heat Warning is issued when there is risk of heat stress to people.

We'll take a look at the different types of illnesses that one can contract from the heat and the sun in just a moment, but let's first look at how heat affects the body.

# Overheating

Humans' blood sits at a fairly constant 98.6 degrees if all is well. But when it gets hot, we pant and we sweat. The heart beats a bit harder, pumping more blood through the dilating blood vessels and into the capillaries near the skin where the heat is transferred.

> **Eye of the Storm**
>
> Sunburn actually can significantly retard the body's ability to shed excess heat.

Once near the skin's surface, the heat drains into the cooler air and water pushes through the pores to form a cooler layer.

Heat illnesses happen when that sweat meets humid air and isn't allowed to evaporate, which would cool us, and so we overheat.

## Beat the Heat

Okay, so now we generally know how a hot day and a lawnmower or that extra mai tai can cause some malady, but what exactly are we talking about when we talk about heat and sun illnesses?

In the former, we're talking heat cramps, heat exhaustion, and heatstroke. In the latter, it's sunburn and even skin cancer.

### Heat Cramps

Everybody gets heat cramps—everybody. When it's hot out and the body is really working, tilling that garden or skateboarding down to the park, those muscles spasm and it feels like Armageddon in your calf or stomach. Even zillion-dollar football stars out on the Dallas gridiron buckle like babies—with all those trainers and high-energy sports drinks around them ….

Your body has too little fluid in it, and your electrolyte content—the sodium, potassium, magnesium, chloride, and calcium levels in your body—is flat-lining. Stretching will help keep your muscles from cramping up, but a constant and moderate intake of a sports drink will replace those electrolytes and keep you on your feet.

Just remember, too much liquid with all that jostling about and you're going to get stomach cramps, so keep it in moderation.

**Storm Stats**

The average person should drink 8 to 12 glasses of water each day.

Remember, the human body is about 67 percent water, and that water delivers oxygen and nutrients to the cells, regulates body temperature, removes waste, and lubes joints.

When the weather's hot and humid, the first thing you may notice about dehydration is that you're thirsty. Seems simple enough. But the fact is that if you're already thirsty, you're already in that down-spiral, so get water or a sports drink quickly.

The trick is to hydrate before and during strenuous activity. A simple rule of thumb is that you must replace the fluids you lost. So if you lost 2 pounds while exercising, consume two to three cups (sixteen to twenty-four ounces) of fluid for every pound.

**Storm Stats**

When exercising in hot, humid environments, the human body can lose 6 pints of fluid an hour.

## Heat Exhaustion

Not drinking enough liquids when you're in the heat also can cause heat exhaustion, more severe than heat cramps.

Heat exhaustion causes folks to get lightheaded, dizzy, nauseated, tired, anxious, and confused or disoriented. It's a dangerous place to be, since you need your mental capacities to tell you to get some water in you and to get to a cooler place.

And that's the first step—get to a cooler place, splash cold water on yourself or get a wet towel, and drink, drink, drink.

If you're confused, feverish, or get lethargic, you need to get someone to take you to the hospital. Your body wants to close up shop, and an IV is probably going to be the cure.

Dehydration means a whole lot of pain. It literally will feel as if your insides are on fire and could cause uncontrollable vomiting or dry heaves. You need medical attention quickly at this point.

### Heatstroke

We often hear about this one, but never really understand it. That's because you may never know you're suffering a heatstroke until it's too late. You may not even be sweating.

But you will be dizzy and confused and your temperature will be very high.

The condition can be life-threatening, so never take it lightly. Have someone get you to the hospital quickly.

Shed some clothing, drink liquids, get into a cool place, and generally treat the condition like you would heat exhaustion, but you must get to the doctor. Serious damage or loss of life could occur.

## Sunshine Go Away?

The sun can seriously burn skin, and prolonged exposure to the sun can cause skin cancer. Don't rely on waiting to feel burned to get out of the sun. Take precautions. Cover up and wear sunscreen, and pay attention to the UV Index.

### Sunburn

First-degree burns are common for Spring Breakers coming from New York to spend a week in the hot sun of Florida. But they are common with most other folks, too. Everybody's gotten sunburned once or twice in his or her life.

But sunburns need to be avoided like the plague. The sun's rays are powerful, and can fry your unprotected exposed skin in as quick as 15 minutes, causing redness and burning in the epidermis—the first layer of skin.

Second-degree burns involve the skin's second layer, or the dermis. Blisters are common in a second-degree burn, which is much more painful than a first-degree burn.

Light clothing and a hat are good ways to keep the effects of the sun at bay, and sun block with a high SPF rating works well when skin is exposed to the sun. Remember to reapply hourly or more, depending on how much you're sweating.

**0-2 UV Index: Low**

- Minimal sun protection is required for normal activity.
- Wear sunglasses on bright days. If outside for more than one hour, cover up and use sunscreen.
- Reflection of snow can nearly double UV strength. Wear sunglasses and apply sunscreen.

**3-5 UV Index: Moderate**

- Take precautions: cover up, and wear a hat, sunglasses, and sunscreen — especially if you will be outside for 30 minutes or more.
- Look for shade near midday when the sun is strongest.

**6-7 UV Index: High**

- Protection is required: UV damages the skin and can cause a sunburn.
- Reduce time in the sun between 11A.M. and 4P.M. and take full precautions (seek shade, cover up and wear a hat, sunglasses, and sunscreen).

**8-10 UV Index: Very high**

- Extra precautions are required: unprotected skin will be damaged and can burn quickly.
- Avoid the sun between 11A.M. and 4P.M. and take full precautions (seek shade, cover up, and wear a hat, sunglasses, and sunscreen).

**11+ UV Index: Very high**

- Take full precautions: unprotected skin will be damaged and can burn in minutes. Avoid the sun between 11A.M. and 4P.M., cover up, and wear a hat, sunglasses, and sunscreen.
- Values of 11 or more are very rare in Canada. However, the UV Index can reach 14 or more in the tropics and southern United States.
- White sand and other bright surfaces reflect UV and increase UV exposure.

*The UV Index.*

*(Courtesy of the National Weather Service.)*

Remember, you can get a sunburn on a cloudy day—the sun's ultraviolet rays penetrate clouds—so cover up with clothing or sunblock.

People who suffer severe sunburns may experience nausea, fever, and chills. They should see a doctor.

For lesser sunburns, a cool bath (no bubble bath, salts, or fragrances, but cooked oatmeal is good) will temporarily stop the burning, and aloe vera will help once you get out of the tub. Aspirin every four hours will help the pain, but stay out of the sun, for Pete's sake.

## Skin Cancer/Melanoma

Long-term overexposure to the sun's radiation can cause melanoma, a skin cancer. Dark-colored tumors or swelling of the skin is usually how it manifests itself. A mole may become darker or bigger. Anytime you see an unusual spot or growth on the skin, you should see a doctor to rule out cancer.

**Storm Stats**

An average of 9,000 people die each year from skin cancer, and in 2002, 53,000 new cases of melanoma were found.

There are two types of skin cancer, nonmelanoma and melanoma. The latter is the more severe of the two. Nonmelanoma occurs in the basal or squamous cells, at the base of the outer layer of skin. Most of these cancers develop on areas most exposed to the sun: face, nose, ears, neck, backs of hands, and the tops of balding heads.

They don't usually spread to other parts of the body, but when they do, they can spread quickly.

Melanoma, on the other hand, can spread to other body parts and can be deadly. The cells that produce melanin give the skin its color or pigment. Melanin protects the deeper layers of the skin from the sun. Melanoma can be prevented when caught early, but if not caught early, it does cause the majority of skin cancer deaths.

If you notice any of the symptoms that we've listed here or a change in the skin's pigment, you need to get checked out by a doctor. Remember, as with any form of cancer, the earlier you catch it, the better your chances of surviving it.

## Animal Logic

Don't think for a second that your pet is cool just because you are. Dogs and cats that spend time outdoors are extremely susceptible to heat and humidity. Their coats can act as insulation against the sun, but it also can cause them problems when they can't cool down. An estimated 5,000 pets die each year due to extreme temperatures, hot and cold, but no figures exist as to how many die from the heat.

However, it's unfortunately not uncommon to see dogs waiting in the car for their owners to return from shopping or ordering a burger.

A car in a parking lot is a lethal place for a pet to be during the warmer months. Car interiors can reach temperatures of 150 degrees within an hour, and that's with the windows rolled down. Windows up, even cracked enough for air, will kill a dog in just a few minutes.

Best advice? Leave Fido and Fluffy in your air-conditioned home while you make the run for that ice cream cone.

### Hot Dogs (and Cats)

Dogs shouldn't be left out in the heat—even their dog houses can reach temperatures that are hot enough to send them into heat stress or heatstroke.

And since dogs can't communicate well, it's important to take care that the weather doesn't do them in.

However, dogs are better at communicating when they are sick than cats. When a dog is suffering from heatstroke, it will pant more than normal and bark or whine. He or she also will have a worried expression on their face and even a furrowed brow.

**Storm Stats**

Short-nosed dogs are more affected by heat stress. Pekingese, pugs, boxers, and bulldogs are particularly at risk.

*Dogs need plenty of shade and water during the summer, especially. Dogs don't sweat, so they need to stay out of the sun for long stretches.*

*(Courtesy of Christopher Passante.)*

If left unrelieved, the dog will pant excessively and will not be able to breathe properly. Eventually, the dog will drool excessively and vomit and ultimately collapse. The gums of the dog will turn blue and he or she will convulse or become unconscious.

Cats are just as much at risk for heatstrokes as dogs. And more at risk than humans.

Like dogs, cats cool themselves by breathing hard. The don't sweat, but they lick themselves to emulate sweat. The saliva evaporates and cools them. But if the temperature is too hot, the cat can overheat.

Again, like dogs, short-nosed cats are more susceptible to heatstroke—Persians, especially. And as with any pet, cats shouldn't be left in a hot car or hot crate.

Cats that are suffering from heatstroke will exhibit a bright red tongue, vomiting, staggering, diarrhea (sometimes with blood), pale blue or gray lips, and unconsciousness.

Without help, the animal will die.

## Prevention and Treatment

For both dogs and cats, prevention is the best measure. Keep your cat and/or dog indoors when it's hot outside. Always have fresh, cool water for them. Be sure it can't tip over and don't put the water in the sun.

*Keep cats indoors on hot days. Cats will lick themselves to create the cooling sensation of sweat.*

*(Courtesy of Christopher Passante.)*

If the pet is outdoors, be sure to have cool water there, too, as well as a shady area where the dog or cat can retreat to.

Never leave the dog or cat confined to a shed or a hot house for long periods of time.

If your cat or dog will be outdoors in the summer, you can clip the long-haired species to keep them cool. Watch out for sunburn, though. Cats and dogs both can suffer burns and develop skin cancers.

Exercise your dogs in the morning or evening, when the air isn't too hot. Running on a hot day can kill a dog.

If the cat or dog is showing signs of heat stress or heatstroke, bring them to a cooler temperature and call your veterinarian immediately.

You can immerse the pet in cool water, hose them off, or wrap them in cool, wet towels.

# Dangers of a Hotter Planet

The more we understand our weather, climate, and planet's natural characteristics, the better we can cope with elements like heat and humidity.

As we discussed in Chapter 2, the global warming debate continues to heat up, but many experts believe that we're already seeing the adverse health effects of a warmer planet.

The experts include World Health Organization scientists who believe that 160,000 people die from global warming effects every year and that by 2020, those numbers will have doubled. The associated diseases range from malnutrition to malaria.

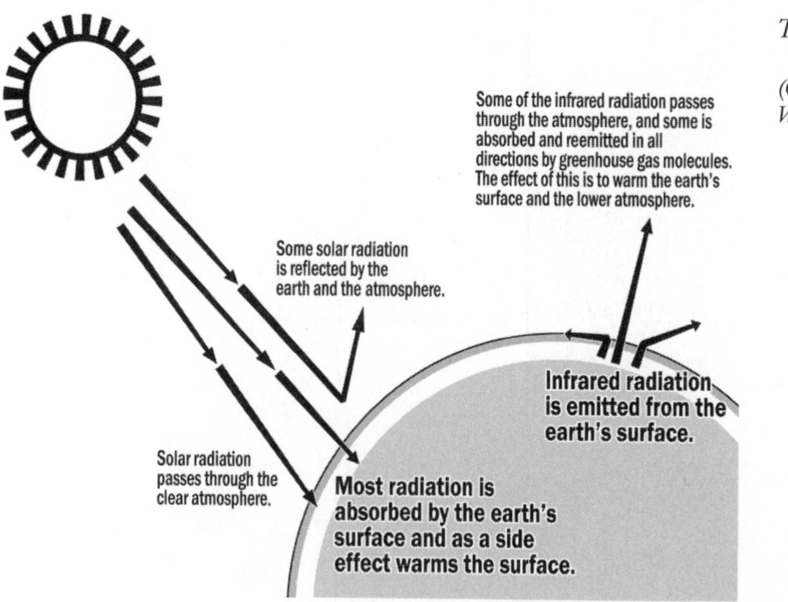


*The greenhouse effect.*

*(Courtesy of the National Weather Service.)*

Small shifts in temperature could extend the range of malaria-spreading mosquitoes, for instance, or water contamination from floods could occur, not to mention soil and farmland erosion.

More frequent and more severe storms also play a part in the global warming equation, but what about the effects of heat on humans?

Well, some scientists believe that more heat waves will occur, and more deaths will result. Consider this: 50,000 Europeans died in a heat wave in 2003. If global warming raises the temperatures—especially the night temperatures—enough, more of these heat waves could occur. Model projections estimate that climate-related diseases will more than double by 2030 and that heat-related deaths in California alone could double by 2100.

## The Least You Need to Know

- Heat waves wreak havoc on humans and animals each year.
- Taking proper precautions in heat can save your life.
- Pets are extremely susceptible to heat.
- Global warming might lead to more disease and death.

Chapter 21

# Hail of a Storm

### In This Chapter

- Hail to thee
- Seeing it coming
- Extreme hailstorms
- Hail alley

If people were to walk outside and see rain "falling" backward—from the ground to the sky—they'd probably make their peace, thinking that the end of the world was upon them.

Then again, while gazing up at the backward rain, the last thing they may ever see on this planet is the grapefruit-sized chunk of ice bearing down on them. End of *their* world, anyway.

That chunk of ice was hail, and it forms when rain "falls" backward.

## What Is Hail?

Hail forms high in the clouds, where water is supercooled enough to freeze.

The thing is that since hail falls during thunderstorms, and thunderstorms primarily happen when it's warm out, it doesn't freeze on the way down, but on the way up.

But by the time it makes it to earth, it actually melts, and we say "Man, those are some big raindrops!"

*A grapefruit-sized hailstone is actually several hailstones fused together and called an aggregate hailstone.*

*(Courtesy of the National Severe Storms Laboratory.)*

They are big raindrops—or small ones—but they may have been hail moments before. Lucky you.

We're not always that lucky, however. And to count our blessings, let's get a clearer picture.

## How Hail Forms

Thunderstorms, as we've learned, are powerful beasts with some weird tics. One of them is the updraft. Warm air is being sucked up to fuel the storm. The wind from that updraft can take the rain that was falling down back up into the clouds to freeze it. When it's heavy enough, it returns to earth in these aerodynamic white particles.

**Inside the Storm**

Sleet is rain that freezes on its way down; freezing rain only freezes when it hits the cold ground; and hail freezes when winds cause an updraft, sending the rain back up into the sky.

Extreme hailstones—hailstones that are as big as eggs, baseballs, or even grapefruits—are whipped around in the clouds longer. They rise and fall and clash and fuse together. If you've ever seen a grapefruit-sized hailstone, you can actually see the pattern of several chunks of hail that have fused together.

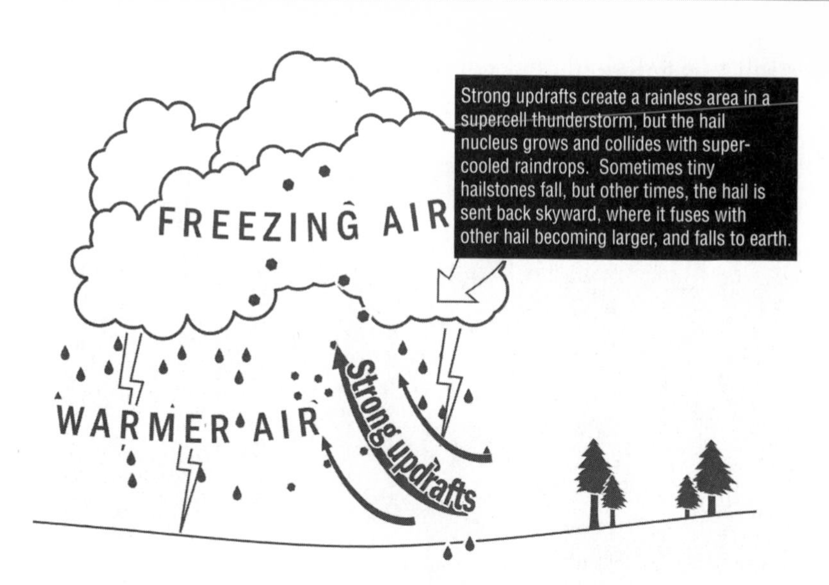


*How hail forms.*

*(Courtesy of the National Weather Service.)*

The bigger they get, the harder they fall. And the more damage they do. We'll check out some extreme hailstorms in a little bit, but for now let's say the sky's the limit on how large a hailstone can become. The wind speeds inside a thunderstorm can reach 120 miles per hour. Imagine a baseball in a 120-mile-an-hour wind and you get the idea.

That baseball's going to whirl around in those clouds until it's good and ready—read: heavy enough—to beat the updraft and let gravity do its thing.

> **Eye of the Storm**
>
> A Hastings, Nebraska, record hailstone had a circumference of 18.75 inches. Although barely smaller, a Coffeyville, Kansas, hailstone weighed 1.65 pounds.

## Severe Hail

Severe hail is defined as having a diameter of 0.75 inches or more. That's about the size of a jellybean, but weather people say dime-sized. These things won't exactly injure you, unless they accumulate and you slip or skid on them.

They don't feel great, though. But the big-daddy hail, those hailstones that are the size of a navel orange, can destroy roofs and homes, total cars, and kill animals—and people.

These extreme hailstones form in supercell thunderstorms.

If you remember from Chapter 6, a supercell thunderstorm is a thunderstorm that has a deep, persistent rotating updraft called a mesocyclone.

These storms have incredible updrafts and can keep hail aloft for some time, as they grow and grow and grow ….

*Smaller hailstones band together to form large hailstones. This one is about 4 inches in diameter.*

*(Courtesy of the National Severe Storms Laboratory.)*

# Hail Indicators

It stands to reason that some of the worst hail forms in some of the strongest thunderstorms. And with the strongest thunderstorms usually including tornadoes, you can imagine the kind of hail one of those mesocyclones delivers.

The funny thing is that during a tornado not too many folks are focusing on the hail.

Legend has it that hail precedes a tornado. Don't bank on that. While hail will indicate that there is a thunderstorm present, and large hailstones will tell you that this mesocyclone has extremely powerful updrafts, it doesn't mean there's a tornado on the horizon. And most importantly, just because there isn't hail falling from the sky doesn't mean there isn't a tornado on the radar.

**Storm Stats**

One of the most costly thunderstorms to ever hit the Dallas-Fort Worth, Texas, area caused as much as $2 billion worth of damage in May 1995, much of that because of hail.

Hail usually forms in storms with high cloud tops that can range between 20,000 feet and 30,000 feet or higher. That's where the freezing layer of the storm usually is. So, the cloud tops have to reach high enough into the freezing layer to get hail formation. The stronger the updraft, the higher the cloud top, the stronger the storm, and the greater the likelihood of hail formation.

*A severe thunderstorm delivers an array of hail that accumulates in sizes up to 3 inches.*

*(Courtesy of the National Severe Storms Laboratory.)*

But hailstorms don't last all that long. Usually, within a few minutes they've come and gone. That's because the updrafts don't last that long—maybe 10 to 15 minutes. The rain-cooled air weakens the updraft and puts an end to hail development.

Meteorologists can forecast hailstorms. They look for the possibility of severe storms to form. Some key items to look for are heavy rain, gusty and damaging winds of at least 58 miles per hour or greater, different winds with height—termed wind shear—high cloud tops ranging between 20,000 and 30,000 feet or greater, supercell storm possibilities, and tornado potential.

**Eye of the Storm**

That hailstorms don't last that long is no consolation to farmers whose entire fields of crops have been destroyed by hail in just a few minutes.

# Worst Hailstorms in History

It's hard to measure the very worst hailstorm in world history. As noted, hail is usually accompanied by wind, rain, lightning, and even tornadoes. The combined intensity of a storm like this could yield heavy damages and casualties. Independently, each of these facets of the storm can be deadly and damaging.

Still, if you had to pick a very deadly hailstorm, you wouldn't have to look much farther than China on July 19, 2002, for one of the worst hail events to ever happen.

**Storm Stats**

A hailstorm in Munich, Germany, in 1984 damaged more than 70,000 buildings and 250,000 cars, and injured 400 people.

*Golf-ball-sized hail from a storm near Roosevelt, Oklahoma, on April 9, 1978.*

*(Courtesy of the National Severe Storms Laboratory.)*

A prelude to a thunderstorm on this summer day, hailstones the size of eggs began falling in the central Henan province of China and folks couldn't run for cover fast enough.

**Storm Stats**

Hailstone damage averages about $1 billion each year in the United States alone.

Twenty-two people were killed on that day and an estimated 200 were injured. Winds picked up, toppling buildings, trees, and a cattle factory.

Back in the United States, a hailstorm in May of 1995 packed a pretty good wallop. In fact, it was the worst-ever severe thunderstorm in North America.

The date was May 5, and like most severe thunderstorms, forecasters knew this one was Texas-sized. What they couldn't know was that in its wake, an estimated $2 billion in damages would occur.

**Eye of the Storm**

Hail can fall at high speeds. The hail that fell during the Mayfest storm of 1995 reached 80 miles per hour.

It was so bad that the second-worst storm, a July 1990 hailstorm in Denver, Colorado, caused a "mere" $625 million in damages with its softball-sized hail.

More than 500 people were injured when the Texas storm began dropping hail on the greater Dallas area during a Mayfest celebration that more than 100,000 people had attended. When the hail began falling, there was nowhere to go. People took refuge in their cars or huddled together as the hail streaked to earth at speeds estimated at 80 miles per hour. No one was killed during the event, but nearly 100 people were injured.

Death came later as the storm continued through the night.

In its wake, 13 people were killed and cars, homes, crops, and businesses suffered great losses.

The Mayfest storm, as with most major thunderstorms, included softball-sized hail that pummeled the area, flash floods, and 70-mile-per-hour winds.

**Storm Stats**

The most deadly hailstorm in world history killed 246 people and 1,600 domesticated animals in India on April 30, 1888.

The storm was part of a supercell that moved into the Dallas-Fort Worth area overnight on May 5 and 6.

*New cars are damaged at a car dealership in Amarillo, Texas, after a hailstorm dropped baseball-sized hail on June 21, 2004.*

*(Courtesy of the National Weather Service.)*

The wind uprooted trees and toppled buildings. The hail left holes in roofs and windows of homes and destroyed many brand-new cars in car dealership lots. A General Motors plant alone sustained $20 million in damages, and as many as 20 RVs in a park were piled up against a fence after high winds forced them there.

After the hail and wind, the rains came hard. Flash flooding made the streets of Dallas look like lakes with waist-deep water. At least 16 people drowned.

The hail was so intense that the ground looked as if it had been covered in snow. Any trees and bushes in its way were stripped bare.

## Where the Hail Is

When we talk about hailstorms, we tend to think they happen only once in a great while. But for many people in certain areas of the world, it's a lot more common than one might think.

**Storm Stats**

Northeast India tends to have the largest hailstones in the world.

As with any type of storm—blizzards in Rochester, New York, lightning in Tampa, Florida, tornadoes in Oklahoma City—there are many areas prone to hail, and weather folks are well-versed in being able to pass out this information.

They call it Hail Alley, and it is the region where Wyoming, Nebraska, and Colorado meet, just east of the Rocky Mountains. It hails here more than anywhere else in the United States.

In fact, the average is seven to nine hail days a year. Most of it will fall during late afternoons in late spring.

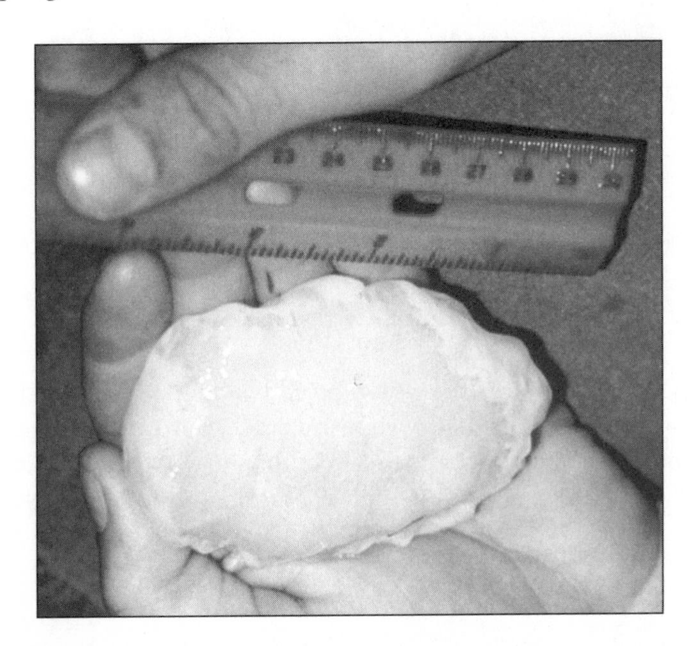

*Baseball-sized hail was collected during a severe thunderstorm on June 8, 1995, in Texas.*

*(Courtesy of the National Severe Storm Laboratory.)*

Now, that might not seem like a whole lot of hail, but consider this: how many times do you see hail each year where you live?

## The Least You Need to Know

- Hail is formed when rain is caught in an updraft, freezes, and falls.
- Hail can kill people and damage crops in just seconds.
- The worst American hailstorm was in Dallas-Fort Worth, Texas, on May 5 and 6, 1995.
- An area just east of the Rockies is home to the most frequent hailstorms in the United States.

Appendix A

# Glossary

**accessory cloud** A cloud that depends on bigger clouds to develop.

**advection** Atmospheric property movement from the wind.

**air-mass thunderstorm** A thunderstorm that has no association to a front. Air mass thunderstorms are created in warm, humid air during the afternoon, and dissipate after dusk. These storms rarely are as severe as other thunderstorms.

**anticyclonic rotation** Movement opposite the earth's rotation—clockwise in the Northern Hemisphere.

**anvil** When a cumulonimbus cloud becomes flat at the top, and takes the shape of an anvil.

**anvil dome** A large overshooting top above an anvil.

**approaching** A thunderstorm containing wind of 40 to 57 miles per hour, or hail from ½ inch to ¾ inch in diameter.

**arcus** On the leading edge of a thunderstorm, this is a low, horizontal cloud formation.

**back-building thunderstorm** When the new development of a thunderstorm is on the upwind side, so the storm looks like it's not moving or is moving backward.

**backing winds** Winds shifting or changing counterclockwise in direction and height.

**blizzard** Snow combined with winds of at least 35 miles per hour and visibility of less than a quarter-mile for a period of three hours or more.

**boundary layer** A layer of air beside a bounding surface.

**bow echo** A linear radar echo bent in a bow shape.

**bubble high** An area of high pressure in a mesoscale that is associated with cooler air from the rainy downdraft area of a thunderstorm.

**cap** A warm air level in the atmosphere that stunts thunderstorm development by bringing in cooler air to “cap” the storm formation.

**cell** Convection as a single updraft, downdraft, or updraft-downdraft combination that looks like a dome in a cumulus cloud.

**cirrus** High clouds containing ice and appearing in the form of white filaments or patches in narrow, sweeping, hairlike bands.

**clear slot** Clearing skies or reduced cloud cover that indicate drier air within a cloud base.

**closed low** A low-pressure area with a distinct cyclonic circulation center that can be surrounded by isobars or height contour lines.

**cloud streets** Rows of cumulus clouds that are aligned with low-level flow.

**cold advection** Cold air moving into a region by horizontal winds.

**cold-air funnel** A funnel cloud developed from a small shower or thunderstorm when the air above is unusually cold.

**collar cloud** A circular cloud ring surrounding the upper part of a wall cloud.

**comma cloud** A cloud pattern that is comma-shaped, usually associated with large and intense low-pressure systems.

**condensation** Water vapor in the air that turns into liquid water. The opposite is evaporation.

**condensation funnel** A funnel-shaped cloud consisting of condensed water droplets.

**confluence** Wind-flow pattern where air moves inward toward an axis that is parallel to the general direction of flow. It is the opposite of difluence.

**congestus** Towering cumulus.

**convection** The movement of heat and moisture by the movement of a fluid, specifically during vertical transport of heat and moisture by updrafts and downdrafts in an unstable atmosphere.

**convective temperature** Approximate temperature that the air near the ground must warm to for convection on the surface to develop.

**cumuliform anvil** A thunderstorm anvil that looks like cumulus-type clouds and rises from rapid spreading of a thunderstorm updraft.

**cumulus** Dense independent clouds that show vertical development such as domes, mounds, or towers, with round tops and flat bases.

**cutoff low** A closed low that is cut off from basic westerly currents and is independent.

**cyclic storm** A thunderstorm that weakens and grows in intensity while maintaining its individuality.

**cyclogenesis** Development of a low-pressure center, such as a cyclone.

**cyclonic rotation** The counterclockwise rotation of a system in the Northern Hemisphere.

**derecho** Convection-based windstorm that moves rapidly, such as downburst clusters, and produces winds over areas more than 100 miles wide and long.

**dew point** Atmospheric moisture measurement. The temperature that air must be cooled to reach saturation.

**difluence** A wind-flow pattern where air fans out away from a central axis that is parallel to the general direction of the flow. Opposite of confluence.

**diurnal** Daily.

**divergence** The spreading out of a vector field, such as in horizontal winds. Opposite of convergence.

**Doppler radar** Radar measuring radial velocity, the instantaneous component of motion parallel to the radar beam.

**downburst** A strong downdraft of wind generally associated with thunderstorms.

**downdraft** A small air column that sinks rapidly toward the ground, generally joined by precipitation such as a thunderstorm.

**dry line** A boundary separating moist and dry air masses.

**dry line storm** A thunderstorm that develops near a dry line.

**dust devil** A small vortex independent of a thunderstorm that rotates with dust and debris and formed by surface heating during hot weather.

**erosion** The abrasional wearing away of the ground or other surface by liquid (water) or air.

**evaporation** How water becomes water vapor, which includes vaporization from water surfaces, land surfaces, and snow fields.

**evapotranspiration** The product of evaporation and transpiration.

**extratropical** A cyclone that has lost its "tropical" characteristics.

**F Scale** See Fujita Scale.

**feeder bands** Lines or bands of low-level clouds that feed into the updraft region of a thunderstorm, usually from the east through the south.

**flood** An overflow of water onto land not normally covered by water.

**flood, 100-year** A flood level with a 1 percent chance of being equaled or exceeded in any given year.

**flood stage** The level of overflow on banks of a stream or water body.

**floodplain** A strip of relatively flat and normally dry land alongside a stream, river, or lake that is covered by water during a flood.

**freezing rain** Rain that falls onto a frozen surface, causing the rain to freeze.

**front** A boundary between two different air masses and temperature.

**Fujita Scale** (or F Scale) A scale of wind damage intensity:

F0: 40–72 mph, light damage.
F1: 73–112 mph, moderate damage.
F2: 113–157 mph, considerable damage.
F3: 158–206 mph, severe damage.
F4: 207–260 mph, devastating damage.
F5: 261–318 mph, (rare) incredible damage.

**funnel cloud** A rotating funnel that extends from a cumulus cloud downward, but does not touch the ground, differing from a tornado.

**glacier** An ice mass that forms on land by the compaction of snow and moves slowly down or outward by its own weight.

**ground water** Water that flows or seeps downward and saturates soil or rock, supplying springs and wells.

**humidity** The amount of the water vapor content of the air.

**hurricane** An intense cyclonic tropical storm with winds greater than 74 miles per hour with heavy wind and rain, moving in a clockwise or counterclockwise direction, depending on which hemisphere it starts in.

**hurricane season** In the Atlantic, Caribbean, and Gulf of Mexico, the period from June through November; in the eastern Pacific, June through November 15; and in the central Pacific, June through October.

**hydrologic cycle** The cyclic transfer of water vapor from the earth's surface into the atmosphere, from the atmosphere via precipitation back to earth, and through runoff into streams, rivers, and lakes, and ultimately into the oceans.

**isobar** A line connecting points of equal pressure.

**jet stream** The strong winds of a narrow atmospheric stream in high-altitude winds.

**levee** A natural or human-made barrier on the edge of a stream, lake, or river that protects from flooding.

**mammatus clouds** Rounded, smooth protrusions beneath a thunderstorm anvil.

**medium range** In forecasting, three to seven days in advance.

**mesocyclone** A storm region of rotation measuring approximately 2 to 6 miles in diameter and often found in the right rear of a supercell.

**microburst** A short-lived, small downburst that affects an area under 2.5 miles across.

**mid-level cooling** Air cooling in middle levels of the atmosphere—8,000 to 25,000 feet—which can destabilize the atmosphere if all other factors are equal.

**moisture advection** Moisture transport by horizontal winds.

**monsoon** Seasonal winds caused by the alternate heating and cooling of land and sea. The monsoon is more common in Asia but the Southern and desert regions of the United States, for instance, also have monsoon seasons.

**multiple-vortex tornado** A tornado that occurs when two or more condensation funnels or debris clouds are concurrently present.

**n-point source pollution** Pollution released over a wide area, not from one specific location.

**NOAA** National Oceanographic and Atmospheric Administration.

**nowcast** A short-term weather forecast, generally fewer than six hours.

**NSSFC** National Severe Storms Forecast Center, in Kansas City, Missouri.

**NSSL** National Severe Storms Laboratory, in Norman, Oklahoma.

**NWS** National Weather Service.

**occluded mesocyclone** A mesocyclone that forms when air from the rear-flank downdraft envelopes the circulation at low levels and cuts off the inflow of warm, unstable low-level air.

**orographic lift** Air that is lifted as it moves up and over mountains or slopes.

**outflow boundary** The boundary of a storm-scale or mesoscale that separates thunderstorm-cooled air from the surrounding air.

**overshooting top** Domelike protrusion on top of a thunderstorm anvil, that shows there is a very strong updraft and means there is a higher potential for severe weather.

**percolation** Water movement through the openings in rock or soil.

**permeability** The ability of a material to allow liquid to pass through.

**point-source pollution** Water pollution coming from a single point, such as a sewage-outflow pipe.

**precipitation** Rain, snow, hail, sleet, dew, and frost.

**rear flank downdraft** A dry-air region behind and around a mesocyclone.

**relative humidity** The ratio of atmospheric moisture present relative to the amount that would be present if the air were saturated. Measured in a percentage.

**reservoir** A natural or man-made pond, lake, or basin that stores, regulates, and controls water.

**ridge** A long area of relatively high atmospheric pressure.

**rope funnel** A narrow condensation funnel typically associated with a decaying tornado.

**runoff** Precipitation, snow melt, or irrigation water that appears in uncontrolled surface streams, rivers, drains, or sewers.

**severe thunderstorm** A thunderstorm producing tornadoes, hail 0.75 inches or more in diameter, or winds of 58 miles per hour or more.

**shear** The variation in wind speed or direction over a short distance.

**sleet** Rain droplets that freeze into ice crystals before hitting the ground.

**snowpack** The amount of snow and ice on the ground, including new snow and existing snow and ice.

**squall line** A solid line or band of active thunderstorms.

**Storm Prediction Center** A national forecast center in Norman, Oklahoma, that provides short-term forecast guidance for severe convection, excessive rainfall (flash flooding), and severe winter weather in the United States.

**stratiform** Clouds that have extensive horizontal development.

**stratocumulus** Low-level clouds in a flat layer and arranged in rows, bands, or waves.

**stratus** Low gray cloud layers with a fairly uniform base.

**striations** Parallel grooves or channels in cloud formations.

**subsidence** Sinking motion in the atmosphere, usually over a broad area.

**supercell** A thunderstorm with a strong rotating updraft responsible for tornadoes, large hail, and damaging winds.

**tornado** A strong rotating column of air usually touching both the ground and the base of the thunderstorm.

**towering cumulus** Large cumulus clouds with vertical development, mushrooming up at the top.

**transpiration** When water is absorbed by plants, usually through the roots, then evaporates into the atmosphere from the plant surface.

**tropical cyclone** A strong, organized low-pressure circulation system that begins over tropical or subtropical waters.

**trough** A line of relatively low atmospheric pressure.

**updraft** A small-scale current of rising air.

**upper level system** Any large-scale or mesoscale disturbance that can produce lift in the middle or upper parts of the atmosphere.

**upslope flow** Air that is forced up higher terrain.

**veering winds** Winds that shift in a clockwise direction.

**virga** Streaks of precipitation falling from a cloud but evaporating before reaching the ground.

**warning** A National Weather Service bulletin showing that a particular weather hazard is either imminent or has been reported. A warning indicates the need to take action to protect life and property, according to NWS.

**watch** A National Weather Service bulletin showing that a particular hazard is possible; that conditions are more favorable than usual for its occurrence. A watch is a recommendation for planning, preparation, and increased awareness, according to NWS.

**water cycle** The cycle of water movement from the oceans to the atmosphere and to the earth and returning to the atmosphere through various stages or processes such as precipitation, interception, runoff, infiltration, percolation, storage, evaporation, and transportation.

**water quality** The chemical, physical, and biological characteristics of water, usually in respect to its suitability for a particular purpose.

**water table** The top of the water surface in the saturated part of an aquifer, or underground water source.

**watershed** An area of land that drains water to a particular stream, river, or lake.

**waterspout** A small weak rotating column of air over water beneath a cumulonimbus or towering cumulus cloud. It looks like a tornado over water.

**wind chill** The cooling effect of temperature and wind on exposed skin.

**Winter Storm Warning** A weather bulletin that is issued when hazardous winter weather in the form of heavy snow, heavy freezing rain, or heavy sleet is imminent or occurring within the next 12 hours.

**Winter Storm Watch** A weather bulletin that is issued when there is strong possibility of a blizzard, heavy snow, heavy freezing rain, or heavy sleet and issued 12 to 36 hours before the beginning of the storm.

# Appendix B

# Weather Extremes

## Climate Extremes

### Wettest Places

| Rank | Average Rainfall | Place | Years of record as of 2006 |
|---|---|---|---|
| 1 | 523.6 | Lloro, Colombia | 29 |
| 2 | 467.4 | Mawsynram, India | 38 |
| 3 | 460 | Kauai, Hawaii | 30 |
| 4 | 405 | Debunscha, Cameroon | 32 |
| 5 | 354 | Quibdo, Colombia | 16 |
| 6 | 340 | Queensland, Australia | 09 |
| 7 | 256 | Henderson Lake, Canada | 14 |
| 8 | 183 | Crkvica, Bosnia-Hercegovina | 22 |

## Driest Places

| Rank | Average Rainfall | Place | Years of record as of 2006 |
|---|---|---|---|
| 1 | 0.03 | Arica, Chile | 59 |
| 2 | 0.1 | Wadi Halfa, Sudan | 39 |
| 3 | 0.8 | Amundsen-Scott, Antarctica | 10 |
| 4 | 1.2 | Batagues, Mexico | 14 |
| 5 | 1.8 | Aden, Yemen | 50 |
| 6 | 4.05 | Mulka, Australia | 42 |
| 7 | 6.4 | Astrakhan, Russia | 25 |
| 8 | 8.93 | Puako, Hawaii | 13 |

## Highest Temperature Extremes

| Rank | Temp (F) | Place | Date |
|---|---|---|---|
| 1 | 136 | El Azizia, Libya | 13 Sep 1922 |
| 2 | 134 | Death Valley, Calif. | 10 Jul 1913 |
| 3 | 129 | Tirat Tsvi, Israel | 22 Jun 1942 |
| 4 | 128 | Cloncurry, Australia | 16 Jan 1889 |
| 5 | 122 | Seville, Spain | 04 Aug 1881 |
| 6 | 120 | Rivadavia, Agentina | 11 Dec 1905 |
| 7 | 108 | Tuguergari, Phillipines | 29 Apr 1912 |

## Lowest Temperature Extremes

| Rank | Temp (F) | Place | Date |
|---|---|---|---|
| 1 | -129 | Vostok, Antarctica | 21 Jul 1983 |
| 2 | -90 | Oimekon, Russia | 06 Feb 1933 |
| 3 | -90 | Verkhoyansk, Russia | 07 Feb 1892 |
| 4 | -87 | Northice, Greenland | 09 Jan 1954 |
| 5 | -81.4 | Snag, Canada | 03 Feb 1947 |
| 6 | -69 | Ust'Shchugor, Russia | January (n/a) |

| Rank | Temp (F) | Place | Date |
|---|---|---|---|
| 7 | -27 | Ifrane, Morocco | 01 Jun 1907 |
| 8 | -11 | Charlotte, Australia | 11 Feb 1935 |
| 9 | 12 | Mauna Kea, Hawaii | 29 Jun 1994 |

# Worst Tornadoes

## Most Destructive U.S. Tornadoes, Ranked

| Rank | Date | Location | Deaths |
|---|---|---|---|
| 1 | 18 Mar 1925 | MO/IL/IN | 695 |
| 2 | 06 May 1840 | Natchez, MS | 317 |
| 3 | 27 May 1896 | St. Louis, MO | 255 |
| 4 | 05 Apr 1936 | Tupelo, MS | 216 |
| 5 | 06 Apr 1936 | Gainesville, GA | 203 |
| 6 | 09 Apr 1947 | Woodward, OK | 181 |
| 7 | 24 Apr 1908 | Amite, LA; Purvis, MS | 143 |
| 8 | 12 Jun 1899 | New Richmond, WI | 117 |
| 9 | 08 Jun 1953 | Flint, MI | 115 |
| 10 | 11 May 1953 | Waco, TX | 114 |
| 10 | 18 May 1902 | Goliad, TX | 114 |
| 12 | 23 Mar 1913 | Omaha, NE | 103 |
| 13 | 26 May 1917 | Mattoon, IL | 101 |
| 14 | 23 Jun 1944 | Shinnston, WV | 100 |
| 15 | 18 Apr 1880 | Marshfield, MO | 99 |
| 16 | 01 Jun 1903 | Gainesville, GA | 98 |
| 16 | 09 May 1927 | Poplar Bluff, MO | 98 |
| 18 | 10 May 1905 | Snyder, OK | 97 |
| 19 | 24 Apr 1908 | Natchez, MS | 91 |
| 20 | 09 Jun 1953 | Worcester, MA | 90 |
| 21 | 20 Apr 1920 | Starkville, MS | 88 |

*continues*

### Most Destructive U.S. Tornadoes, Ranked (continued)

| Rank | Date | Location | Deaths |
|---|---|---|---|
| 22 | 28 Jun 1924 | Sandusky, OH | 85 |
| 23 | 25 May 1955 | Udall, KS | 80 |
| 24 | 29 Sep 1927 | St. Louis, MO | 79 |
| 25 | 27 Mar 1890 | Louisville, KY | 76 |

## Worst Hurricanes

### Strongest Hurricanes, Ranked (Above Cat 4) Pressure

| Rank | Hurricane | Location | Year | Cat | (mb) |
|---|---|---|---|---|---|
| 1 | Unnamed | FL | 1935 | 5 | 892 |
| 2 | Camille | MS, LA, VA | 1969 | 5 | 909 |
| 3 | Katrina | MS, LA, AL, FL | 2005 | 4 | 920 |
| 4 | Andrew | FL, LA | 1992 | 5 | 922 |
| 5 | Unnamed | TX | 1886 | 4 | 925 |
| 6 | Unnamed | FL | 1919 | 4 | 927 |
| 7 | Unnamed | FL | 1928 | 4 | 929 |
| 8 | Donna | FL | 1960 | 4 | 930 |
| 9 | Unnamed | LA | 1915 | 4 | 931 |
| 9 | Carla | TX | 1961 | 4 | 931 |
| 11 | Unnamed | LA | 1856 | 4 | 934 |
| 11 | Hugo | SC | 1989 | 4 | 934 |
| 13 | Unnamed | FL, MS, AL | 1926 | 4 | 935 |
| 14 | Unnamed | TX | 1900 | 4 | 936 |
| 15 | Unnamed | GA, FL | 1898 | 4 | 938 |
| 15 | Hazel | SC, NC | 1954 | 4 | 938 |
| 17 | Unnamed | FL, LA, MS | 1947 | 4 | 940 |
| 18 | Unnamed | TX | 1932 | 4 | 941 |
| 18 | Charley | Eastern U.S. | 2004 | 4 | 941 |
| 20 | Gloria | Eastern U.S. | 1985 | 4 | 942 |

## Top 25 Deadliest Hurricanes Est.

| Rank | Name | Year | Location | Deaths |
|---|---|---|---|---|
| 1 | Galveston | 1900 | TX | 8000 |
| 2 | Unnamed | 1928 | FL | 2500 |
| 3 | Katrina | 2005 | MS, LA, FL, AL | 1400 |
| 4 | Unnamed | 1893 | LA | 1100–1400 |
| 5 | Sea Island | 1893 | SC, GA | 1000–2000 |
| 6 | Unnamed | 1881 | SC, GA | 700 |
| 7 | Unnamed | 1935 | FL | 408 |
| 8 | Unnamed | 1856 | LA | 400 |
| 9 | Audrey | 1957 | LA, TX | 390 |
| 10 | Unnamed | 1926 | FS, MS, AS | 372 |
| 11 | Unnamed | 1909 | LA | 350 |
| 12 | Unnamed | 1919 | FL, TX | 287 |
| 13 | Unnamed | 1915 | LA | 275 |
| 13 | Unnamed | 1915 | TX | 275 |
| 15 | Unnamed | 1938 | New England | 256 |
| 15 | Camille | 1969 | MS, LA, VA | 256 |
| 17 | Diane | 1955 | NE US | 184 |
| 18 | Unnamed | 1898 | GA, SC, NC | 179 |
| 19 | Unnamed | 1875 | TX | 176 |
| 20 | Unnamed | 1906 | FL | 164 |
| 21 | Unnamed | 1886 | TX | 150 |
| 22 | Unnamed | 1906 | MS, AL, FL | 134 |
| 23 | Unnamed | 1896 | FL, GA, SC | 130 |
| 24 | Agnes | 1972 | FL, NE US | 122 |
| 25 | Rita | 2005 | TX, LA | 100 |

## Costliest U.S. Hurricanes (unadjusted)

| Rank | Name | State(s) | Year | Cat | Damage |
|---|---|---|---|---|---|
| 1 | Katrina | MS, AL, LA | 2005 | 4 | 80B |
| 2 | Andrew | FL, LA | 1992 | 5 | 26.5B |
| 3 | Charley | FL | 2004 | 4 | 15B |
| 4 | Ivan | AL, FL | 2004 | 3 | 14.2B |
| 5 | Rita | TX, LA | 2005 | 3 | 9.4B |
| 6 | Frances | FL | 2004 | 2 | 8.9B |
| 7 | Hugo | SC | 1989 | 4 | 7B |
| 8 | Jeanne | FL | 2004 | 3 | 6.9B |
| 9 | Allison | TX | 2001 | TS | 5B |
| 10 | Floyd | Mid-Atlantic | 1999 | 2 | 4.5B |
| 11 | Isabel | Mid-Atlantic | 2003 | 2 | 3.3B |
| 12 | Fran | NC | 1996 | 3 | 3.2B |
| 13 | Opal | FL, AL | 1995 | 3 | 3B |
| 14 | Frederic | AL, MS | 1979 | 3 | 2.3B |
| 15 | Agnes | FL, NE U.S. | 1972 | 1 | 2.1B |
| 16 | Alicia | TX | 1983 | 3 | 2B |
| 17 | Bob | NC, NE U.S. | 1991 | 2 | 1.5B |
| 18 | Juan | LA | 1985 | 1 | 1.5B |
| 19 | Camille | MS, LA, VA | 1969 | 5 | 1.4B |
| 20 | Betsy | FL, LA | 1965 | 3 | 1.4B |
| 21 | Elena | MS, AL, FL | 1985 | 3 | 1.2B |
| 22 | Gloria | FL, MS, AL | 1985 | 3 | 9M |
| 23 | Lili | SC, LA | 2002 | 1 | 8.6M |
| 24 | Diane | NE U.S. | 1955 | 1 | 8.3M |
| 25 | Bonnie | NC, VA | 1998 | 2 | 7.2M |

# Storm Names

Note: If hurricane storm names are exhausted in one season, names will follow the Greek alphabet. They are the same for every year: Alpha, Beta, Gamma, Delta, Epsilon, Zeta, Eta, Theta, Iota, Kappa, Lambda, Mu, Nu, Xi, Omicron, Pi, Rho, Sigma, Tau, Upsilon, Phi, Chi, Psi, Omega.

## Atlantic Hurricane Names

| 2006 | 2007 | 2008 | 2009 | 2010 | 2011 |
|---|---|---|---|---|---|
| Alberto | Andrea | Arthur | Ana | Alex | Arlene |
| Beryl | Barry | Bertha | Bill | Bonnie | Bret |
| Chris | Chantal | Cristobal | Claudette | Colin | Cindy |
| Debby | Dean | Dolly | Danny | Danielle | Dennis |
| Ernesto | Erin | Edouard | Erika | Earl | Emily |
| Florence | Felix | Fay | Fred | Fiona | Franklin |
| Gordon | Gabrielle | Gustav | Grace | Gaston | Gert |
| Helene | Humberto | Hanna | Henri | Hermine | Harvey |
| Isaac | Iris | Isidore | Ida | Igor | Irene |
| Joyce | Jerry | Josephine | Joaquin | Julia | Jose |
| Kirk | Karen | Kyle | Kate | Karl | Katrina |
| Leslie | Lorenzo | Lili | Larry | Lisa | Lee |
| Michael | Michelle | Marco | Mindy | Matthew | Maria |
| Nadine | Noel | Nana | Nicholas | Nicole | Nate |
| Oscar | Olga | Omar | Odette | Otto | Ophelia |
| Patty | Pablo | Paloma | Peter | Paula | Philippe |
| Rafael | Rebekah | Rene | Rose | Richard | Rita |
| Sandy | Sebastien | Sally | Sam | Shary | Stan |
| Tony | Tanya | Teddy | Teresa | Tomas | Tammy |
| Valerie | Van | Vicky | Victor | Virginie | Vince |
| William | Wendy | Wilfred | Wanda | Walter | Wilma |

## Retired Atlantic Hurricane Names (Alphabetical)

|  |  |
|---|---|
| Agnes | 1972 |
| Alicia | 1983 |
| Allen | 1980 |
| Allison | 2001 |
| Andrew | 1992 |
| Anita | 1977 |
| Audrey | 1957 |
| Betsy | 1965 |
| Beulah | 1967 |
| Bob | 1991 |
| Camille | 1969 |
| Carla | 1961 |
| Carmen | 1974 |
| Carol | 1965 |
| Celia | 1970 |
| Cesar | 1996 |
| Charley | 2004 |
| Cleo | 1964 |
| Connie | 1955 |
| David | 1979 |
| Diana | 1990 |
| Diane | 1955 |
| Donna | 1960 |
| Dora | 1964 |
| Edna | 1968 |
| Elena | 1985 |
| Eloise | 1975 |
| Fabian | 2003 |
| Fifi | 1974 |
| Flora | 1963 |
| Floyd | 1999 |
| Fran | 1996 |
| Frances | 2004 |
| Frederic | 1979 |
| Georges | 1998 |
| Gilbert | 1988 |
| Gloria | 1985 |
| Gracie | 1959 |
| Hattie | 1961 |
| Hazel | 1954 |
| Hilda | 1964 |
| Hortense | 1996 |
| Hugo | 1989 |
| Inez | 1966 |
| Ione | 1955 |
| Iris | 2001 |
| Isabel | 2003 |
| Isidore | 2002 |
| Ivan | 2004 |
| Janet | 1955 |
| Jeanne | 2004 |
| Joan | 1988 |
| Juan | 2003 |
| Keith | 2000 |
| Klaus | 1990 |
| Lenny | 1999 |
| Lili | 2002 |
| Luis | 1995 |
| Marilyn | 1995 |
| Michelle | 2001 |
| Mitch | 1998 |
| Opal | 1995 |
| Roxanne | 1995 |

## Eastern North Pacific Names

| 2006 | 2007 | 2008 | 2009 | 2010 | 2011 |
|---|---|---|---|---|---|
| Aletta | Adolph | Alma | Andres | Agatha | Adrian |
| Bud | Barbara | Boris | Blanca | Blas | Beatriz |
| Carlotta | Cosme | Cristina | Carlos | Celia | Calvin |
| Daniel | Dalilia | Douglas | Dolores | Darby | Dora |
| Emilia | Erick | Elida | Enrique | Estelle | Eugene |
| Fabio | Flossie | Fausto | Felicia | Frank | Fernanda |
| Gilma | Gil | Genevieve | Guillermo | Georgette | Greg |
| Hector | Henriette | Hernan | Hilda | Howard | Hilary |
| Ileana | Israel | Iselle | Ignacio | Isis | Irwin |
| John | Juliette | Julio | Jimena | Javier | Jova |
| Kristy | Kiko | Kenna | Kevin | Kay | Kenneth |
| Lane | Lorena | Lowell | Linda | Lester | Lidia |
| Miriam | Manuel | Marie | Marty | Madelime | Max |
| Norman | Narda | Norbert | Nora | Newton | Norma |
| Olivia | Octave | Odile | Olaf | Orlene | Otis |
| Paul | Priscilla | Polo | Patricia | Paine | Pilar |
| Rosa | Raymond | Rachel | Rick | Roslyn | Ramon |
| Sergio | Sonia | Simon | Sandra | Seymour | Selma |
| Tara | Tico | Trudy | Terry | Tina | Todd |
| Vicente | Velma | Vance | Vivian | Virgil | Veronica |
| Willa | Wallis | Winnie | Waldo | Winifred | Wiley |
| Xavier | Xina | Xavier | Xina | Xavier | Xina |
| Yolanda | York | Yolanda | York | Yolanda | York |
| Zeke | Zelda | Zeke | Zelda | Zeke | Zelda |

## Central North Pacific Names

Names are used until exhausted.

| | |
|---|---|
| Akoni | Alika |
| Ema | Ele |
| Hana | Huko |
| Io | Ioke |
| Keli | Kika |
| Lala | Lana |
| Moke | Maka |
| Nele | Neki |
| Oka | Oleka |
| Peke | Peni |
| Uleki | Ulia |
| Wila | Wali |
| Aka | Ana |
| Ekeka | Ela |
| Hali | Halola |
| Iolana | Iune |
| Keoni | Kimo |
| Li | Loke |
| Mele | Malia |
| Nona | Niala |
| Oliwa | Oko |
| Paka | Pali |
| Upana | Ulika |
| Wene | Walaka |

## Western North Pacific Ocean/South China Sea

Names are used until exhausted.

Damrey
Longwang
Kirogi
Kai-Tak
Tembin
Bolaven
Chanchu
Jelawat
Ewiniar
Bilis
Kaemi
Prapiroon
Maria
Saomai
Bopha
Wukong
Sonamu
Shanshan
Yagi
Xangsane
Bebinca
Rumbia
Soulik
Cimaron
Chebi
Durian
Utor
Trami
Kong-Rey
Yutu
Toraji
Man-Yi
Usagi
Pabuk
Wutip
Sepat
Fitow
Danas
Nari
Wipha
Francisco
Lekima
Krosa
Haiyan
Podul
Lingling
Kajiki
Faxai

*continues*

*continued*

Peipah
Tapah
Mitag
Hagibis
Noguri
Rammasun
Matmo
Halong
Nakri
Fengshen
Kalmaegi
Fung-Wong
Kammuri
Phanfone
Vongfong
Nuri
Sinlaku
Hagupit
Changmi
Mekkhala
Higos
Bavi
Maysak
Haishen

Pongsona
Yanyan
Kujira
Chan-Hom
Linfa
Nangka
Soudelor
Molave
Koni
Morakot
Etau
Vamco
Krovanh
Dujuan
Maemi
Choi-Wan
Koppu
Ketsana
Parma
Melor
Nepartak
Lupit
Sudal
Nida

Omais
Conson
Chanthu
Dianmu
Mindulle
Tingting
Kompasu
Namtheun
Malou
Meranti
Rananim
Malakas
Megi
Chaba
Aere
Songda
Sarika
Haima
Meari
Ma-On
Tokage
Nock-Ten

Muifa
Merbok
Nanmadol
Talas
Noru
Kulap
Roke
Sonca
Nesat
Haitang
Nalgae
Banyan
Washi
Matsa
Sanvu
Mawar
Guchol
Talim
Nabi
Khanun
Vicente
Saola

## Australian Region Names

Names are used until exhausted.

**Western**

Alex
Bessi
Clancy
Dianne
Errol
Fiona
Grant
Harriet
Iggy
Jana
Ken
Linda
Mitchell
Nicky
Oscar
Phoebe
Raymond
Sally
Tim
Vivienne
Willy
Adeline
Bertie
Clare
Daryl
Emma
Floyd
Glenda
Hubert
Isobel
Jacob
Kara
Lee
Melanie
Nicholas
Ophelia
Pancho
Rosie
Selwyn
Tiffany
Victor
Zelia
Alison
Billy
Cathy
Damien
Ellie
Frederic
Gabrielle
Hamish

Ilsa
Joseph
Kirrily
Leon
Marcia
Norman
Olga
Paul
Robyn
Sean
Terri
Vincent
Yvette
Northern
Amelia
Bruno
Coral
Dominic
Esther
Ferdinand
Gretel
Hector
Irma
Jake
Kay
Laurence
Marian
Neville

Olwyn
Phil
Rachel
Samuel
Thelma
Verdun
Winsome
Alistair
Bonnie
Craig
Debbie
Evan
Farrah
George
Helen
Ira
Jasmine
Kim
Laura
Matt
Narelle
Oswald
Penny
Russell
Sandra
Trevor
Valerie
Warwick

**Eastern**

Alfred
Blanch
Caleb
Denise
Ernie
Frances
Greg
Hilda
Ivan
Joyce
Kelvin
Liz
Marcus
Nora
Owen
Polly
Richard
Sadie
Theo
Verity
Wallace
Alice
Bruce
Cecily
Dennis
Edna
Fletcher
Gillian
Harold
Ita
Jack
Kitty
Les
May
Nathan
Olinda
Pete
Ruby
Stan
Tammie
Vaughan
Wylva
Anika
Bernie
Claudia
Des
Erica
Fritz
Grace
Harvey
Ingrid
Jim

- Kate
- Larry
- Monica
- Nelson
- Odette
- Pierre
- Rebecca
- Sherly
- Tania
- Vernon
- Wendy

## Port Moresby, Papua New Guinea Names

Names are used until exhausted.

- Epi
- Guba
- Ila
- Kama
- Matere
- Rowe
- Tako
- Upia
- Abul
- Emau
- Gule
- Igo
- Kamit
- Tiogo
- Ume

## Fiji Region Names

Names are used until exhausted.

- Ami
- Beni
- Cillia
- Dovi
- Eseta
- Fili
- Gina
- Heta
- Ivy
- Judy
- Kerry
- Lola

*continues*

*continued*

| | |
|---|---|
| Meena | Oli |
| Nancy | Pat |
| Olaf | Rene |
| Percy | Sarah |
| Rae | Tomas |
| Sheila | Vania |
| Tam | Wilma |
| Vaianu | Yasi |
| Wati | Zaka |
| Xavier | Atu |
| Yani | Bune |
| Zita | Cyril |
| Arthur | Daphne |
| Becky | Evan |
| Cliff | Freda |
| Daman | Garry |
| Elisa | Helene |
| Funa | Ian |
| Gene | June |
| Hettie | Koko |
| Innis | Lusi |
| Joni | Mike |
| Ken | Nute |
| Lin | Odile |
| Mick | Pam |
| Nisha | Reuben |

Solo
Tui
Victor
Winston
Yalo
Zena
Amos
Bart
Cora
Dani
Ella
Frank
Gita
Hali
Iris
Jo
Kim
Leo
Mona
Neil
Oma
Paula
Rita
Sose
Trina
Vicky
Waka
Yolande
Zoe
Ana
Bina
Chris
Donna
Eva
Fanny
Glen
Hagar
Irene
Julie
Kala
Louise
MalNate
Olo
Pami
Rex
Tino
Vanessa
Yvonne
Zazu

## Southwest Indian Ocean Names

| 2005–2006 | 2006–2007 | 2007–2008 |
|---|---|---|
| Alvin | Anita | Ariel |
| Boloetse | Bondo | Bongwe |
| Carina | Clovis | Celina |
| Diwa | Dora | Dama |
| Elia | Enok | Elnus |
| Farda | Favio | Fame |
| Geduza | Gamede | Gula |
| Helio | Humba | Hondo |
| Isabella | Indlada | Ivan |
| Jaone | Jaya | Jokwe |
| Kundai | Katse | Kamba |
| Lindsay | Lisebo | Lola |
| Marinda | Magoma | Marabe |
| Nadety | Newa | Nungu |
| Otile | Olipa | Ofelia |
| Pindile | Panda | Pulane |
| Quincy | Quincy | Qoli |
| Rugare | Rabeca | Rossana |
| Sebina | Shyra | Sama |
| Timba | Tsholo | Tuma |
| Usta | Unokubi | Uzale |
| Velo | Vuyane | Vongai |
| Wilby | Warura | Warona |
| Xanda | Xylo | Xina |
| Yuri | Yone | Yamba |
| Zoelle | Zouleha | Zefa |

## North Indian Ocean

Names are used until exhausted.

| | |
|---|---|
| Onil | Chapala |
| Agni | Megh |
| Hibaru | Vaali |
| Pyarr | Kyant |
| Baaz | Nada |
| Fanoos | Vardah |
| Mala | Sama |
| Mukda | Mora |
| Helen | Nisha |
| Lehar | Bijli |
| Madi | Aila |
| Na-nauk | Phyan |
| Hudhud | Ward |
| Nilofar | Laila |
| Priya | Bandu |
| Komen | Phet |
| Ogni | Ockhi |
| Akash | Sagar |
| Gonu | Baazu |
| Yemyin | Daye |
| Sidr | Luban |
| Nargis | Titli |
| Abe | Das |
| Khai-Muk | Phethai |

*continues*

*continued*

| | |
|---|---|
| Giri | Fani |
| Jal | Vayu |
| Keila | Hikaa |
| Thane | Kyarr |
| Mujan | Maha |
| Nilam | Bulbul |
| Mahasen | Soba |
| Phailin | Amphan |

Appendix C

# Further Reading

The following list contains not only the sources used in this book but some fantastic sites and publications you can use to learn more about extreme weather around the world.

AAAS Atlas of Population and Environment
http//atlas.aaas.org
This site shows humans' effects on our world.

Argonne National Laboratory
www.newton.dep.anl.gov/askasci/wea00.htm
This site allows you to search weather facts.

Atlantic Oceanographic and Meteorological Laboratory
www.aoml.noaa.gov/hrd/tcfaq/tcfaqHED.html.
Hurricanes, typhoons, and cyclonic storms data can be found here, along with FAQs.

Audubon
www.audubon.org
Everything nature and natural, with great research and writing.

Australian Government Bureau of Meteorology
www.bom.gov.au
A look at weather down under.

BBC
www.bbc.co.uk
Excellent weather archives, world weather events, and data.

Caroll, Chris. "In Hot Water." *National Geographic.* August 2005. One of the foremost in-depth magazines covering the natural world and society. This edition deals with Florida hurricanes and a warmer planet.

Climate Ark
www.climateark.org
Great blog and site for global warming views and climate change discussion.

CNN
www.cnn.com
Excellent weather archives, world weather events, and data.

Cooperative Institute for Mesoscale Meteorological Studies
www.cimms.ou.edu
One of the foremost clearinghouses for tornado data and research.

Cyberwest Magazine
www.cyberwest.com
An online magazine that deals with weather topics.

Digital Snow Museum
www.wintercenter.homestead.com/photoindex.html
Good weather photos and data by event.

The Disaster Center
www.disastercenter.com
Online reference and clearinghouse on all things weather disaster. Also a chat room.

EUMETSAT
www.eumetsat.int
Searchable site that monitors weather, climate, and environment around the world.

European Environment Agency
www.eea.eu.int
Good environmental data from Europe.

Extreme Science
www.extremescience.com/weatherport.htm
Great site—and lots of fun—to "feed your brain" with weather data.

Federal Emergency Management Agency
www.fema.gov
Check out one of the most comprehensive sites on weather disasters ever assembled.

Guinness World Records
www.guinnessworldrecords.com
Weather extremes abound here.

Hawaii Solar Astronomy
www.solar.ifa.hawaii.edu
Great data on solar- and weather-related topics.

India Regional Meteorological Centre
www.education.vsnl.com/imdchennai
India's National Weather Service.

Japan Meteorological Research Institute
http://www.mri-jma.go.jp/Information/logomark-sjis.html
Japan's National Weather Service.

Live Weather Images
www.weatherimages.org
Cool website to check out weather images.

Meteorological Service of Canada
www.weatheroffice.ec.gc.ca
Weather data from Canada.

National Aeronautics and Space Administration Earth Observatory
www.earthobservatory.nasa.gov/Library/Hurricanes
Very neat government site to see hurricanes and data.

National Climatic Data Center
www.ncdc.noaa.gov
Government site that details climate data.

National Hurricane Center
www.nhc.noaa.gov
The king of hurricane information.

National Lightning Safety Institute
www.lightningsafety.com
The best site to discover more on lighting and thunderstorms.

National Oceanic and Atmospheric Administration
www.noaa.gov
The foremost weather data clearinghouse in the world.

National Oceanic and Atmospheric Administration's National Coastal Data Development Center
www.ncddc.noaa.gov
Very good weather data on this site.

National Oceanic and Atmospheric Administration's National Weather Service Storm Prediction Center
www.spc.noaa.gov
Good data and storm history here.

National Parks Service
www.nps.gov/interp/nasa/hurricane.html
A wealth of information, especially on climate.

National Snow and Ice Data Center
www.nsidc.org
All things cold, snow, and ice; this is the leader.

National Weather Service
www.nws.gov
Along with NOAA, the National Weather Service specifically looks at all things weather and its data is first-rate.

National Weather Service Northwest River Forecast Center
www.nwrfc.noaa.gov
Local branch of NWS.

National Weather Service Southern Regional Headquarters
www.srh.noaa.gov
Local branch of NWS; good hurricane data.

New Scientist.com
www.newscientist.com/home.ns
Interesting site on science, including weather.

*New York Times*
www.nytimes.com
Good archives on historic weather events as well as weather knowledge.

*Old Farmer's Almanac*
www.almanac.com
The very first forecaster, and still one of the most accurate.

Pennsylvania Department of Heath
http://www.dsf.health.state.pa.us/health
Good snow data on this site.

Population Reference Bureau
www.prb.org
To know more about population migration, this is the site.

Southeast Regional Climate Center
www.sercc.com
Hurricanes and warm-weather storms data can be found here.

Spacedaily
www.spacedaily.com
An absolute must-visit site, this has as much to do with what's around us as it does where we live.

Straight Dope
www.straightdope.com
Funny little message board that dispels a few myths. Very interesting site.

Tom Loffman's Sacramento Weather
www.tloffman.com
Sacramento meteorologist with a wealth of great information.

Typhoon2000.com
www.Typhoon2000.ph
Philippines weather service website.

United Nations World Food Programme
www.wfp.org/english
Very interesting site on how the UN deals with hunger, which is many times caused by drought or weather.

United States Census Bureau
www.quickfacts.census.gov
Population data clearinghouse.

United States Coast Guard
www.uscg.mil/USCG.shtm
Very solid weather data and info on flying into hurricanes here.

United States Department of Agriculture
www.usda.gov/wps/portal/usdahome
Good information on how storms affect agriculture, including historical data.

United States Environmental Protection Agency Global Warming Site
www.yosemite.epa.gov/oar/globalwarming.nsf/content/index.html
Very strong and recent data supporting global warming debates.

United States Search and Rescue Task Force
www.ussartf.org
Winter tips on how to survive the extreme cold. Good site.

University of Nebraska-Lincoln School of Natural Resources
www./snrs.unl.edu
Solid weather, especially tornadoes, here.

*USA Today*
www.usatoday.com/money/economy/2005-09-09-katrina-damage_x.htm
Some of the best Hurricane Katrina coverage.

*USA Today*
www.usatoday.com/weather/resources/askjack/archives-hurricane-safety.htm
"Ask Jack" database; very good weather explainers here.

weather.com
www.weather.com
All-purpose weather site.

Weather Doctor
www.islandnet.com/~see/weather/doctor.htm
Keith Heidorn is a weather god. A pretty good artist, too. Very thankful for his contributions to the weather world.

Weather Underground
www.weatherunderground.com
Very solid and interesting weather data site.

WeatherWise
www.profhorn.aos.wisc.edu/wxwise/
Incredible weather-teaching site.

Wikipedia
www.wikipedia.org
Always a dose of perspective.

World Meteorological Organization
www.wmo.ch/index-en.html
Global weather, including global warming.

# Index

## Numbers



## A



## B

## C

## G

## H

## I



## J



## K–L

## M



## N

## O



## P-Q

## R

## S

## T

## U-V



## W-X-Y-Z